NATURALLY OCCURRING ANTIOXIDANTS

Richard A. Larson
Department of Natural Resources and Environmental Sciences
University of Illinois
Urbana, IL

LEWIS PUBLISHERS

Boca Raton   New York

Library of Congress Cataloging-in-Publication Data

Larson, Richard A.
    Naturally occurring antioxidants / Richard A. Larson.
       p. cm.
    Includes bibliographical references and index.
    ISBN 0-87371-957-3 (alk. paper)
      1. Antioxidants. I. Title.
RB170.L37 1997
616.07—dc21   97-8169
                                                                                                                          CIP

This book contains information obtained from authentic and highly regarded sources. Reprinted material is quoted with permission, and sources are indicated. A wide variety of references are listed. Reasonable efforts have been made to publish reliable data and information, but the author and the publisher cannot assume responsibility for the validity of all materials or for the consequences of their use.

Neither this book nor any part may be reproduced or transmitted in any form or by any means, electronic or mechanical, including photocopying, microfilming, and recording, or by any information storage or retrieval system, without prior permission in writing from the publisher.

The consent of CRC Press LLC does not extend to copying for general distribution, for promotion, for creating new works, or for resale. Specific permission must be obtained in writing from CRC Press LLC for such copying.

Direct all inquiries to CRC Press LLC, 2000 Corporate Blvd., N.W., Boca Raton, Florida 33431.

© 1997 by CRC Press LLC
Lewis Publishers is an imprint of CRC Press LLC

No claim to original U.S. Government works
International Standard Book Number 0-87371-957-3
Library of Congress Card Number 97-8169
Printed in the United States of America 1 2 3 4 5 6 7 8 9 0
Printed on acid-free paper

Dedication

To Robert Williams Tuveson (1931–1992)

Preface

Although details are sketchy, evidence points to the probability that life on our planet originated roughly 4 billion years ago, in the absence of atmospheric oxygen. The first life forms that produced oxygen, perhaps as early as 3 billion years ago, may have been able to use the abundant solar energy (possibly even including short-wave ultraviolet radiation) to reduce carbon dioxide and, in the process, to excrete molecular O_2 as a waste product. For a very long time, this oxygen must have been rapidly consumed by reduced substances such as ferrous iron, which geological evidence indicates was abundant in the primeval waters and surface rocks of the planet. However, as these organisms prospered, this gradual titration of reduced species ended, and the excess oxygen began to enter the atmosphere. Although the rate of increase of oxygen concentration is still being debated, several lines of evidence suggest that over a geologically brief period of perhaps 30 million years, it changed from perhaps one-tenth of the current level to roughly what it is today. At the same time, photochemically synthesized ozone began to be produced.

Both these developments presented life with unprecedented hazards. Never before had potent oxidants been present in such significant concentrations in the immediate vicinity of living tissue. There must have been tremendous selective forces that favored the perhaps rare forms that were able to overcome the increasingly toxic environment of the era.

For reasons that will be summarized in this book, the evolution of naturally occurring antioxidants must have been an important factor that permitted life to continue and to prosper. These protective agents fall into two broad and overlapping categories. The first category includes catalytic antioxidants — enzymes such as catalase, superoxide dismutase, and peroxidases — that are able to effectively deactivate potentially hazardous oxidizing agents and, in some cases, even make use of them for the cell's own purposes. This class of antioxidants will not be addressed in this book since they have been so extensively studied and reviewed by biochemists over the past 40+ years.

A second class of antioxidants are what might be termed stoichiometric antioxidants. These are usually small molecules that the cell can rapidly produce and transport into regions where oxidative stress is taking place. There has been growing interest in these compounds in the last couple of decades, driven by several related developments in other areas of science. For example, in the plant sciences, oxidative stress has been recognized to be one of several stress factors such as drought and thermal effects important for the growth and development of plants. Also, in food science, antioxidants have long been recognized as important preservatives, but the potential dangers of synthetic additives have focused new interest on the use of naturally occurring chemicals to perform this function.

In this book, I have tried to combine a description of the more important classes of these compounds with a more general view of the process of autooxidation and antioxidation. I believe that if students of naturally occurring antioxidants keep in mind some of the quantitative measures that are available to assess their effectiveness, we can approach a more complete understanding of important substances and their reactions.

Today's environment, like that of some primeval eras, represents a period of rapid change, largely induced by humanity's massive combustion and utilization of fossil fuels. The oxidation products of these fuels, such as CO_2, enter the atmosphere to a degree where significant changes in the radiative budget near the planet's surface may be taking place (global warming). At the same time, other synthetic compounds, including chlorofluorocarbons have drifted into the stratosphere, where they are entering into catalytic reactions that may change the concentration of ozone, permitting more damaging ultraviolet radiation to reach the earth's surface. We cannot know what these processes, occurring in tandem, may mean for life on earth in the future. We also cannot predict what other significant environmental stresses may be found. Nevertheless, it will continue to be important to have methods, hopefully quantitative ones, that will allow us to address the truly important environmental and health questions of our time.

This book is the product of a love affair with natural products that has lasted nearly 40 years. I count myself lucky to have been able to spend so much time investigating their properties and functions, and to have uncovered so many interesting facts about nature. However, the book would not have been possible had it not been for the extraordinary teachers and collaborators I have had the good fortune to encounter. Out of many such individuals I can mention only a few: Ray Dodson and Ed Leete from the University of Minnesota, Alan Battersby and Ron Parry from Cambridge University, Tom Bott and James Mears from the Academy of Natural Sciences of Philadelphia, Tom Mabry and Linda Macmahan from the University of Texas, and May Berenbaum and Karen Marley from the University of Illinois.

Finally, I must pay homage to those people that Borges called "wonderful friends given to me by my reading," the many scientists who have labored to make sense of the confusing world of naturally occurring antioxidants. These individuals have taught me anonymously, but their teaching is no less important for that. Their names will be found among the references.

Rick Larson

*Reality may avoid the obligation to be interesting,
but hypotheses may not* — Borges

Table of Contents

1. **Autooxidizable Substances: Weathering** 1
   I. Oxidation and Autooxidation 1
      A. Nonbiological Materials: Weathering 3
      B. Biotic Materials: Aging 4
         1. Peroxidation of Lipids 5
         2. Protein Oxidation 8
         3. Oxidative Damage to DNA 9
         4. Autooxidation of Other Biomolecules 11
            a. Carbohydrates 11
            b. Vitamins 13
      C. The Measurement of Autooxidation 15
         1. Oxygen Uptake 15
         2. Peroxide Determination (Active Oxygen Method) 15
         3. Formation of Malondialdehyde and Other TBA-Reactive Substances 17
         4. Detection of Other Carbonyl Compounds 18
         5. Conjugated Double Bond Formation in Lipids 19
         6. Aliphatic Hydrocarbon Production in Lipid Oxidation 19
         7. Light Emission from Cells, Lipids, and Proteins 20
         8. Kinetic Methods 20

2. **Autooxidation Mechanisms** 25
   I. Free Radical Chain Reactions 25
      A. Nature of Radical Species 25
      B. Formation of Free Radicals (Initiation) 29
      C. Reactions of Initiator Free Radicals with Substrates (Propagation) 31
      D. Destruction of Free Radicals (Termination) 34
   II. Singlet Oxygen-Induced Reactions 38
   III. Other Mechanisms of Oxidative Damage in Cells and Tissues 42
      A. Superoxide 42
      B. Hydrogen Peroxide 44
      C. Hydroxyl Radical 45

3. **Quenching and Scavenging of Reactive Species** 51
   I. Radical Scavenging and Redox Potential 51
      A. Preventive Antioxidation 51
      B. Chemical Antioxidation 52

  II. Metal Complexation and Inhibition of Radical Reactions 53
    A. Amino Acids and Peptides 56
    B. Hydroxy and Polycarboxylic Acids 57
  III. Quenching of Singlet Oxygen 58
    A. Physical Quenching 58
    B. Quenching with Reaction 59
  IV. Destruction of Peroxides 59
  V. Synergistic Mechanisms 61

4. Kinetics and Mechanisms of Inhibited Autoxidation 67
  I. Kinetics of Autoxidation in the Absence of Inhibitors 67
    A. Initiation 68
    B. Propagation 69
    C. Termination 71
    D. Overall Rate Law 72
  II. Autoxidation in the Presence of Inhibitors 73
    A. Inhibition by Preventive Antioxidants 73
      1. Complexing Agents 73
      2. Light Absorbers 74
      3. Peroxide Decomposers 75
    B. General Free Radical Scavengers 75
    C. Inhibition by Alkyl Radical Scavengers 76
    D. Inhibition by Chain-Breaking Antioxidants 76
      1. Competition 76
      2. Diffusion 78

5. Phenolic and Enolic Antioxidants 83
  I. Redox Reactions of Phenols 83
    A. One-Electron Oxidations of Phenols 83
    B. Reactions of Phenoxyl Radicals 84
    C. Phenols as Electron Donors 86
  II. Reactions of Phenols with Oxidizing Free Radicals 87
    A. Reactions with Hydroxyl Radical 87
    B. Reactions with Superoxide 87
    C. Reactions with Peroxyl Radicals 90
  III. Reactions of Phenols with Singlet Oxygen 92
  IV. Classes of Phenolic and Enolic Antioxidants 94
    A. Vitamin E and Related Compounds 94
    B. Vitamin C 98
    C. Flavonoids and Derivatives 100
      1. Flavonoids 100
      2. Isoflavonoids 106
      3. Chalcones and Catechins 108
    D. Phenolic Acids and Derivatives 110
      1. Free Phenolic Acids 110

   2. Phenolic Esters and Amides; Tannins 115
  E. Lignans 119
  F. Cucurmin and Derivatives 121
  G. Hydroquinones and Quinones 122
  H. Other Phenolic Compounds 126

6. Nitrogenous Antioxidants 141
 I. Uric Acid and Other Purines 141
 II. Amino Acid and Peptide Derivatives 143
 III. Alkaloids and Related Compounds 145
 IV. Tetrapyrroles 155
 V. Other Nitrogen Compounds 162

7. Sulfur-Containing Antioxidants 169
 I. Glutathione and Other Amino Acid Derivatives 169
  A. Glutathione 169
  B. Carnosine 172
  C. Ergothioneine 172
  D. Ovothiol 173
 II. Lipoic Acid and Related Sulfides and Polysulfides 174

8. Carotenoids and Related Polyenes 179
 I. β-Carotene 180
 II. Oxygen-Containing Carotenoids 183
 III. Retinol and Derivatives 186

Index 191

NATURALLY OCCURRING ANTIOXIDANTS

1 AUTOOXIDIZABLE SUBSTANCES: WEATHERING

I. OXIDATION AND AUTOOXIDATION

Electron transfer is one of the most fundamental processes in chemistry. The passage of an electron or a pair of electrons from a donor (reducing species) to an acceptor (oxidizing species) results in a change in properties for both parties to the reaction. Oxidation was once defined as the incorporation of oxygen into a substance, but now can be more precisely defined as the conversion of a chemical substance into another having fewer electrons; oxidation, therefore, is the loss of one or more electrons to another substance, which, having gained them, is reduced. The propensity of chemical compounds to undergo reduction or oxidation has been studied for nearly 300 years, probably beginning with the Becher-Stahl theory of combustion, popularly known as the "phlogiston theory," formulated in the early 18th century. Its adherents believed that every combustible substance contained a "principle of fire," phlogiston, that was given up during burning. Oils, for example, burned almost completely and were therefore, in these terms, practically pure phlogiston. The theory was decisively overturned by the end of the century, however, due to Lavoisier's quantitative demonstrations that products of combustion actually weighed more than the starting material, and Priestley's discovery of oxygen (see McCann, 1978).

By the early 1800s, it had become clear that oxygen was intimately involved not only in combustion but in other forms of chemical change. In particular, the oxidative decomposition of organic substances has been studied for more than 175 years (de Saussure, 1820). For a long time, oxygen appeared to become incorporated into the structures of susceptible materials without the necessity of other reagents and without manipulation of other reaction parameters such as temperature. The process was not without elements of mystery; it seemed to be spontaneous, proceeding without any apparent external cause. The term "autoxidation" (or "autooxidation") was adopted as a descriptive expression to delineate these reactions from other oxidative reactions.* In the second half of the 19th century, several authors noted instances in which materials such as rubber, vegetable oils, and turpentine underwent autooxidative deterioration that was accompanied by the uptake of oxygen from the air (see, e.g., Hoffman [1860]). The first citation for the word in the Oxford English Dictionary comes from the year 1883 in a paper from the American journal, *Science*, entitled "Autoxidation in living vegetable cells." In this article, the anonymous author reports on the discovery of readily air-oxidized substances in plants whose oxidation gives rise to hydrogen peroxide, which "can, under the influence of diastase, ... cause further oxidations."

After a long period of discussion and speculation concerning the mechanisms of autooxidation, it was finally recognized in the 1920s and 1930s that it was a radical process; molecular oxygen combined in various ways with organic free radicals (organic substances with one or more unpaired electrons) to form peroxides and related compounds, in which the oxygen of the air became incorporated into the structure of the autooxidizing material. The details of the mechanisms will be presented in Chapter 2. Beginning in the 1940s, antioxidants began to be deliberately added to various materials, beginning with natural substances such as nordihydroguaiaretic acid (1-1) and gum guaiac, and later with petroleum-derived synthetic compounds such as BHA (1-2) and BHT (1-3).

(1-1)

(1-2)

---

* Although the spelling *autoxidation* is preferred in most publications (it is favored by about 13:1 over *autooxidation*, according to a key word search of *Chemical Abstracts* from 1967–1996), the latter will be used in this book on grounds of euphony and analogy with other words beginning with the prefix *auto-* (autoantibody, autoimmune, etc.).

$$\underset{\text{(1-3)}}{\underset{CH_3}{\underset{|}{\bigodot}}\overset{OH}{\overset{|}{\underset{}{}}}}$$

(H$_3$C)$_3$C — [2,6-di-tert-butyl-4-methylphenol structure] — C(CH$_3$)$_3$

(1-3)

A. Nonbiological Materials; Weathering

The Buddha's last words were supposedly "All composite things decay; strive diligently." Virtually all substances made up entirely or in part of organic carbon compounds decompose over time. Wood, plastics, petroleum, leather, paper, paints, waxes, etc., all undergo oxidative decomposition reactions at various rates. A synthetic polymer, for example, can be damaged by oxidation at any time during its existence; during initial polymerization, during fabrication, or during use. Oxidation induces many chemical and physical changes into a product, which may include changes in viscosity, brittleness, discoloration, surface cracking, and loss of impact or tensile strength. If unchecked, the process can lead to complete loss of usefulness of the item.

Oxidative decomposition may be initiated by any number of events. Thermal processes, absorption of gamma rays or high-energy ultraviolet photons, ozone- or metal ion-induced reactions all may contribute in specific instances.

Different substances are differentially susceptible to oxidative damage. Polyethylene, $-(CH_2)_n-$, for example, is practically immortal as far as we know. Only because there are a few impurities and oxidized regions of the polymer introduced during processing is there any chance at all for decay. The polymer really has no good sites for attack by oxidizing reagents of the type found at or near the surface of the earth (oxygen, ozone, hydrogen peroxide, etc.). By contrast, polystyrene, $-(CH_2CHPh)_n-$ is much more susceptible to oxidation because of structural features that include the aromatic rings and the reactive (benzylic) hydrogens localized at the phenyl-substituted carbons. Polystyrene objects exposed to environmental conditions degrade rather rapidly unless they are protected with antioxidants, with a useful life of possibly a year or two.

Rubber (1-4) is a polymer that does not age well. Everyone is familiar with old rubber objects that have cracked and decayed over the years. As a general rule, polymers that contain unsaturated groups, such as rubber, are readily autooxidized. Free radical attack occurs at the $-CH_2-$ groups; and ozone, always present in the atmosphere, reacts

$$(-CH_2-\underset{\underset{CH_3}{|}}{C}=CH-CH_2-)_n$$

(1-4)

readily with the double bonds of the structure. To deter these damaging processes, rubber objects usually contain antioxidants or radical scavengers such as hindered amines, BHT (1-3), and carbon black (also a potent absorber of light), or are "vulcanized" by heating with elemental sulfur, which also has antioxidant properties.

The oxidative weathering of petroleum products has been widely studied as a consequence of concerns over oil spills (Payne and Phillips, 1985). Petroleum oils are highly variable mixtures; crude oils are staggeringly complex mixtures containing hydrocarbons as well as compounds containing oxygen, nitrogen, and sulfur atoms as well as organometallic and inorganic fractions. Refined petroleum products are somewhat simpler but still contain a very large number of compounds. Some refined commodities, such as home heating oil, tend to be rich in two- and three-ring aromatic compounds that strongly absorb sunlight. This leads to the likelihood of light-induced photo-oxidation for such materials. In addition, readily oxidized compounds such as tetrahydronaphthalenes and indans (1-5) are often abundant. Petroleum products often contain added antioxidants and anti-wear additives such as BHT (1-3) or metal thiophosphates.

$$(CH_3)_n$$

(1-5)

B. Biotic Materials: Aging

Living things have colonized a wide range of habitats whose oxygen concentrations vary over the full range possible for the earth's surface and near-surface environments. Therefore, evolution has provided organisms with a range of defense mechanisms for existing in the hazardous oxygenated environment of the earth's surface. However, the defense systems are not perfect, and damage to various constituents of the cell constantly occur, as well as accumulate during aging as the mechanisms for damage control become less and less efficient.

Certain disorders are caused directly by oxidizing free radicals, for example, radiation sickness whose proximate cause is suspected to be hydroxyl radical (HO·) produced by the radiolysis of water inside the cells. The HO· so produced almost randomly attacks a variety of biomolecules including lipids, proteins, and DNA (see Chapter 2-C). In many other acute and chronic diseases of humans, animals, and plants, oxidizing species have been implicated.

From a molecular perspective, it is obvious that cells contain organic and inorganic compounds in a very wide range of oxidation states and with a correspondingly wide range of redox potential. (The oxidation state of an organic carbon atom, in simple terms, reflects the degree to which the atom is substituted by other atoms having greater electronegativity, such as N, O, or halogens. The more bonds from the carbon atom that are associated with electronegative elements, the more oxidized the substance is. A hydrocarbon, that is, a compound having only C–C and C–H bonds, is fully reduced. A simple alcohol or amine with a C–OH or C–NH$_2$ functional group is somewhat oxidized. Further oxidation leads to, for example, a carbonyl group, C=O, where two bonds are associated with oxygen; a carboxylic acid, –COOH, with three such bonds; and finally CO$_2$, which is fully oxidized.)

Because oxidation is intimately involved with metabolism, cellular materials are in a constant state of flux; compounds are introduced into the cell, oxidatively transformed into metabolic energy and into the building blocks of cellular growth, and excreted. In all of these processes, redox changes are taking place.

Intracellular structures have a high degree of organization. For example, cell membranes exist as bilayers of polar lipids whose hydrophilic head-groups are associated with aqueous media. Inside the bilayer, the hydrophobic regions of the lipids interact with one another to maintain the specific configuration of the membrane. Proteins are also present in the membrane; they are currently thought of as floating "islands" in the "sea" of lipids (Figure 1-1). Obviously, such a highly organized structure can be readily disrupted by oxidative reactions that convert the membrane constituents to products having different polarities; and, in fact, membrane damage is one of the more common causes (or consequences: Kappus, 1987; Halliwell et al., 1992) of oxidative degradation in cells.

1. Peroxidation of Lipids

Lipids are highly reduced molecules whose structures all prominently feature aliphatic hydrocarbon moieties in some form or other. Cholesterol

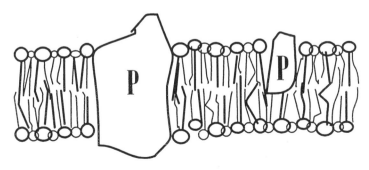

Figure 1-1 Schematic diagram of a lipid bilayer membrane. Circles represent polar head groups of lipid molecules attached to nonpolar hydrocarbon "tails." Objects marked P are protein molecules.

(1-6), for example, is representative of a group of steroidal derivatives featuring cyclohexane and cyclopentane ring systems. Cholesterol is practically a hydrocarbon; only one of its carbon atoms is oxidized, and it only by a single –OH group.

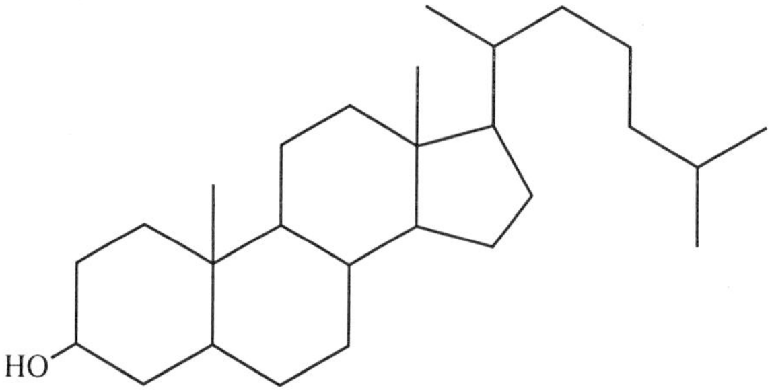

(1-6)

Simple triacylglycerols, or triglycerides, such as tristearin (1-7), are esters of long-chain hydrocarbon (fatty) acids with glycerol. Of the 57 carbon atoms of tristearin, 51 are fully reduced. The situation is similar for more complex lipids. Therefore, all lipids have a high potential to be oxidized.

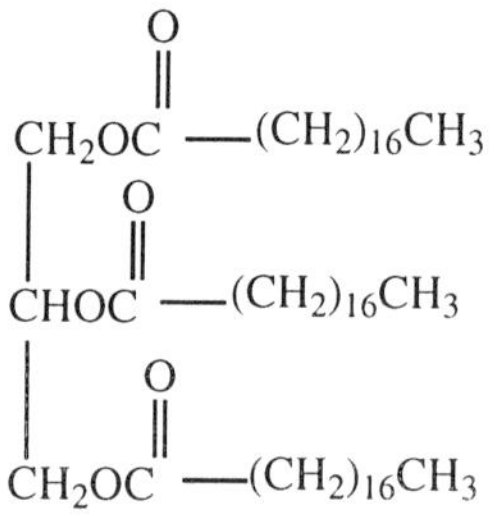

$$\begin{array}{l} CH_2OC(=O)-(CH_2)_{16}CH_3 \\ CHOC(=O)-(CH_2)_{16}CH_3 \\ CH_2OC(=O)-(CH_2)_{16}CH_3 \end{array}$$

(1-7)

The topic of lipid oxidation has aroused great interest in recent years. Historically, lipid oxidation was the provenance of food scientists who were interested in the development of off-flavors and unpleasant odors in stored foodstuffs. In many instances, these problems were shown to be closely correlated with the oxidative deterioration of simple and complex unsaturated fatty acids, whose products had undesirable organoleptic properties and which, furthermore, underwent reactions with other food constituents to produce changes in their colors, flavors, odors, or textures. More recently, lipid oxidation, or peroxidation, has been implicated in chronic and acute disease states in living organisms. Damage and disease in such disparate tissues as liver, lung, and skin have been linked by numerous investigators to membrane or lipid oxidation. The evidence for linking lipid peroxidation products with disease is of two types: first, occurrence of these products within the damaged tissues, and secondly, indications that the isolated products have toxic and mutagenic effects (Kanner et al., 1987).

Lipid peroxidation may be initiated by either one- or two-electron agents. The most well-studied situation is that in which an external oxidant, usually an oxygen-centered free radical such as HO· or ROO·, attacks the lipid chain at a susceptible site such as an allylic methylene group and converts it to a new, carbon-centered free radical (Equation 1-1). This mechanism is typical of the "general" autooxidation process outlined in Chapter 2. Other one-electron oxidants, such as transition metal cations having valence states lying one unit apart (Cu^{2+} and Cu^+, for example), also are capable of initiating many lipid oxidation processes.

$$RCH=CHCH_2 \xrightarrow{HO\cdot} RCH=CH\dot{C}H - + H_2O$$

(Equation 1-1)

Two-electron oxidants include singlet oxygen, discussed more fully in Chapter 2, which with unsaturated lipids undergoes an "ene reaction," leading to the incorporation of oxygen into the lipid chain and migration of a double bond. An example is shown in Equation 1-2.

(Equation 1-2)

Products of lipid peroxidation result when a lipid free radical reacts with molecular oxygen to produce a peroxyl radical, or when a hydroperoxide from singlet oxygen attack rearranges. Subsequent reactions of the peroxyl radical give rise to a variety of oxygenated substances — alcohols, hydroperoxides, ketones, epoxides, etc. Each of these compounds can undergo further reactions, leading to an increasingly complex mixture of products. Crosslinking also occurs, either between lipid molecules of between lipids and proteins; and chain scission of polypeptides also may result (Zirlin and Karel, 1969).

Lipid peroxidation leads to membrane damage because the geometry of the alkyl chains becomes greatly altered and the packing order in the bilayer is disrupted. Membrane damage is sometimes thought of as synonymous with lipid peroxidation. It is beyond question that many lipids, particularly those with multiple double bonds, are readily oxidized by a number of pathways. However, recent findings are casting into question the accepted mechanism,

$$\text{toxic agent} \rightarrow \text{lipid peroxidation} \rightarrow \text{cell damage}$$

Some investigators (cf. Halliwell et al., 1992) believe that in many cases, lipid peroxidation is a secondary consequence of cell damage induced by attacks on DNA or protein targets, and that it may accompany rather than cause cell death. As an example, a significant consequence of membrane damage is leakage of essential metabolites from within the cell. Ca(II), which is an important regulatory agent for protein synthesis, may increase rapidly, leading to cell death. Currently, the belief is that this process is under the control of membrane proteins, and that their damage could lead to significant disruptions of cellular integrity. This would tend to support the belief that protein modification could be a primary mechanism in the initiation of cellular damage. Nevertheless, lipid peroxidation is certainly a real phenomenon that deserves careful attention from those concerned with the complex mechanistic aspects of oxidative stress and toxicity.

2. Protein Oxidation

The amino acids that make up proteins are, in general, more highly oxidized than lipids; each contains a highly oxidized carboxyl group (–COOH) and a partly oxidized amino group ($-NH_2$). The side chains of these molecules, however, usually contain fully reduced carbon atoms as well as other functional groups that, at least theoretically, could undergo oxidative transformations. Several amino acids that occur in proteins have been shown to be especially susceptible to oxidative damage. Whether the oxidant is ozone (Berlett et al., 1991), hydroxyl radical (Davies, 1987), singlet oxygen (Matheson and Lee, 1979), or HOO· (Bielski and Shiue, 1979), the most reactive species appear to be cysteine, histidine, tryptophan, methionine, and phenylalanine, usually in that order. (Arginine and asparagine appear, in addition, to be unusually reactive toward HOO·.)

Free radical production has been implicated in many instances of protein damage (fragmentation, cleavage, and cross-linking). The consequences of these events may include loss of enzyme activity, cytolysis, and even cell death (Wolff et al., 1986). However, in some cases, enzymes such as mixed-function oxidases degrade proteins, presumably as part of the normal cellular activity of protein turnover, using hydrogen peroxide and Fe^{2+} (Fucci et al., 1983). In either case, one of the principal mechanisms of protein damage appears to be the generation of reactive oxygenated free radicals, either HO· or iron-associated forms such as ferryl or perferryl derivatives.

Not as much attention has been given to radical mechanisms of protein damage, relative to lipid and DNA damage, probably because it was believed that damaged protein molecules would not accumulate in cells, but be turned over and removed rapidly. However, more recent evidence suggests that many damaged enzymes accrue within cells as they age, to the point that as much as 30% of the protein in old individuals might consist of oxidized forms (Stadtman, 1992). Protein oxidative damage, as measured by an increase in protein carbonyl content, has been correlated closely with physiological aging in houseflies (Sohal et al., 1993).

Protein damage may occur either in conjunction with or independent of lipid damage. Peroxidizing lipids can, of course, damage proteins that are associated with them in membranes. However, proteins are found in other types of cellular environments having a wide range of properties with respect to the hydrophobicity of their surroundings. Therefore, soluble proteins may be attacked by different classes of oxidants (and be protected by different classes of antioxidants) than membrane proteins. It is possible that protein damage could occur even in an environment where lipid peroxidation was fully protected (Dean et al., 1991). Radical damage to bovine albumin, lysozyme, and mellitin has been demonstrated to take place by, at least in part, a chain mechanism (see Chapter 2); for each reactive free radical generated by a radiolytic method, an average of 15 amino acids were consumed (Neužil et al., 1993). Oxygen was required for the reaction to occur and it could be controlled by chain-breaking antioxidants.

3. Oxidative Damage to DNA

Since it is the most important biopolymer in cells, DNA's chemical reactions have been extensively studied. Radiochemists, in particular, have studied ionizing radiation-induced damage to DNA for approximately 50 years, and have identified many products formed largely by oxidative free radical mechanisms, particularly hydroxyl radicals formed by the radiolysis of water (Ward, 1981). Potential targets for oxidative damage in the DNA chain include the purine and pyrimidine bases as well as the deoxyribose sugar moieties. Specific damage to the bases leaves the strand intact, but modification of the sugar residues may lead to strand breakage (von Sonntag, 1987; Halliwell and Aruoma, 1991).

DNA and its bases can be attacked by all of the principal oxidizing species, but rates of attack vary tremendously. Most investigations of the rate constants of HO· with nucleic acid bases and nucleosides in aqueous solution indicate little discrimination; second-order rate constants are usually in the $4-9 \times 10^9$ l/mol sec range, with guanosine and thymine derivatives sometimes reported as slightly more reactive than the others (Buxton et al., 1988). Initial attack on the bases is almost, but not quite, random, and a few major initial products have been described, including 5-hydroxymethyluracil (1-8) and 8-hydroxyguanine (1-9) (Mouret et al., 1991). Further attack by HO· results in a variety of ring cleavage products (Blakely et al., 1990).

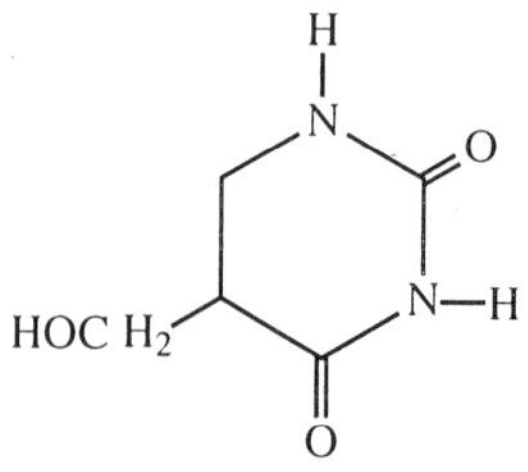

(1-8)

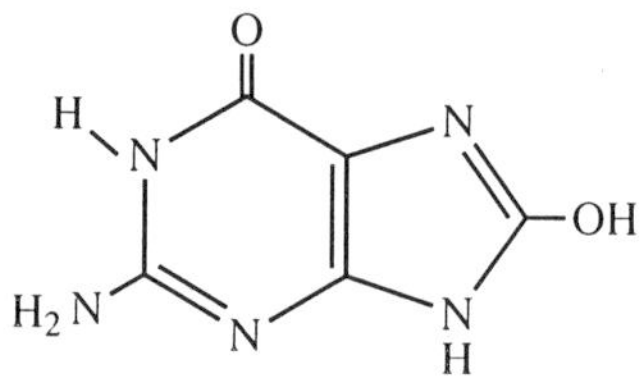

(1-9)

Within the DNA helix, the shielding effect of the sugar phosphate residues probably reduces the extent of attack on the bases by HO·, but data are inconsistent. Hydroxyl radicals are recognized to oxidize DNA at the sugars, which can lead to base liberation, nicking (single-strand) or frank (double-strand) breaks, and interstrand crosslinkage (von Sonntag, 1987).

Hydrogen peroxide may, of course, be a source of strong oxidants such as HO· or related reactive species in the presence of metal ion catalysts such as Fe^{2+} (Fenton reaction; see Chapter 2). This mode of oxidation leads to general damage and strand breakage (Schweitz, 1969) as described above for ·OH. In the absence of free transition metal cations, it has been demonstrated to attack DNA largely at adenine sites, with the product being adenine-N_1-oxide (1-10) (Rhaese, 1968; Mouret et al., 1991).

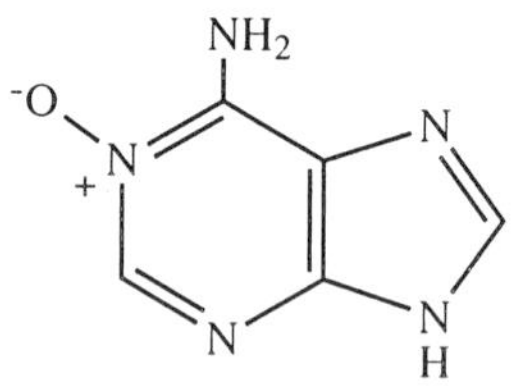

(1-10)

Similarly, hydroperoxides (ROOH) by themselves show little reactivity toward DNA components, but in the presence of transition metal oxidants they are converted to the much more reactive alkoxyl radicals, RO·. DNA and its components have been shown to be highly susceptible to damage under these conditions (Shi et al., 1996; Hazlewood and Davies, 1995).

Carbon-centered peroxyl radicals, ROO· are significantly more selective than hydroxyl or alkoxyl radicals. They do not appear to attack DNA bases (Simic, 1988). However, several types of inorganic peroxyl radicals do appear to have a limited degree of reactivity toward DNA components. For example, sulfonylperoxyl radicals derived from cysteine ($CysSO_2OO·$) selectively abstract a hydrogen from thymine (Razskazovskii and Sevilla, 1996), and also may undergo other reactions with sugar moieties,

Singlet oxygen, a very selective oxidant (see Chapter 2), has been known for many years to attack guanine (1-11) and guanosine residues in solution or in DNA almost exclusively (cf. Gutter et al., 1977; Cadet, 1978; Lee and Rodgers, 1987). This would be expected since guanine is the most electron-rich DNA base. Second-order rate constants for singlet oxygen attack on guanine or deoxyguanosine are in the range of $5-6 \times 10^6$ l/mol sec, which is much lower than for its reaction with other cellular constituents such as β-carotene or the more reactive amino acids (Devasagayam et al., 1991). The products of $^1O_2$ oxidation of guanine derivatives are not fully understood; a number of compounds have been detected using various reaction conditions (Floyd et al., 1989; Piette, 1991). 8-Hydroxyguanine (1-9) is a major product of $^1O_2$ attack (Floyd et al., 1989), and other investigators have also detected it using different oxidants (Shigenaga and Ames, 1991; Shi et al., 1996). Some investigators have reported strand breakage in DNA exposed to sensitizing agents, but the yields are always low and mechanisms other than $^1O_2$ oxidation are hard to rule out.

(1-11)

4. Autooxidation of Other Biomolecules

a. Carbohydrates. Carbohydrate autooxidation has been studied by two major groups; food scientists interested in the chemical changes that occur during the heating of mixtures containing mono- and polysaccharides, and medical scientists concerned with the intracellular carbohydrate-liked reactions that occur during aging and chronic disease states such as diabetes.

Carbohydrate oxidations are usually initiated by one-electron oxidants such as transition metal ions (Cu^{2+}, Fe^{3+}) or oxidizing free radicals (HO·, RO·). Metal

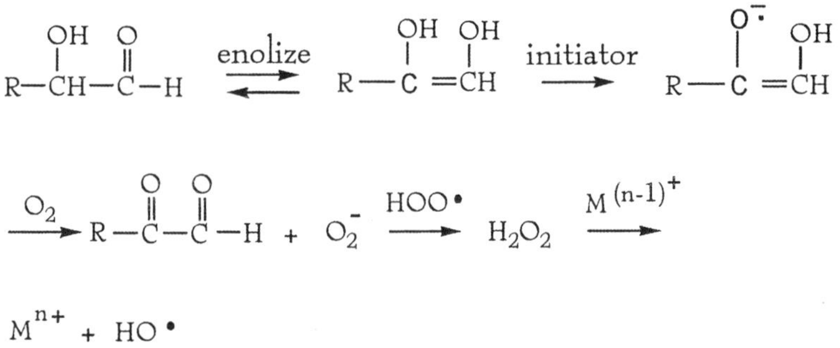

$$R-\underset{\underset{\text{OH}}{|}}{\text{CH}}-\underset{}{\overset{\overset{\text{O}}{\|}}{\text{C}}}-H \underset{\longleftarrow}{\overset{\text{enolize}}{\longrightarrow}} R-\underset{\underset{\text{OH}}{|}}{\text{C}}=\underset{\underset{\text{OH}}{|}}{\text{CH}} \overset{\text{initiator}}{\longrightarrow} R-\underset{\underset{\text{O}^-}{|}}{\text{C}}=\underset{\underset{\text{OH}}{|}}{\text{CH}}$$

$$\overset{O_2}{\longrightarrow} R-\overset{\overset{O}{\|}}{C}-\overset{\overset{O}{\|}}{C}-H + O_2^- \overset{HOO\bullet}{\longrightarrow} H_2O_2 \overset{M^{(n-1)+}}{\longrightarrow}$$

$$M^{n+} + HO\bullet$$

Figure 1-2 Mechanism for carbohydrate autooxidation, as proposed by Wolff and Dean, 1987. (From Wolff, S. P. and R. T. Dean, Biochem. J., 245, 243–250, 1987. With permission.)

ions and oxidants are the principal intermediates in a mechanism proposed by Wolff and Dean (1987) that is supported by kinetic and product analyses (Figure 1-2).

Early oxidation products are α-dicarbonyl compounds, either constrained within the cyclic structure of the sugar (Wolff and Dean, 1987) or cleaved free of it (glyoxal, methylglyoxal, etc.). Reducing sugars can be considered as α-hydroxycarbonyl compounds, which are known to autooxidize from their enol forms by one-electron transfer to oxygen, hydrogen peroxide, or some other oxidant (Wolff et al., 1984: Figure 1-2). Since superoxide and/or hydroxyl radicals are produced during this reaction, it has the potential to proceed by a chain mechanism.

Maillard, or browning, reactions occur when compounds with carbonyl groups (such as reducing sugars or sugar decomposition products) are heated together with compounds having free amino groups (Hodge, 1953; Namiki, 1988). These reactions lead to very complex mixtures of products having a wide range of molecular weights. Among the simple compounds that are formed are furfural (1-12a) and 5-hydroxymethylfurfural (1-12b), which are formally dehydration and rearrangement products of glucose. Hydroxymethylfurfural is the major volatile compound formed during caramelization, which is defined as the thermal decomposition of pure sugars (Defaye and Fernandez, 1995). Many other furan derivatives have also been reported, although usually at lower concentrations. The formation of these compounds is also promoted by the presence of free amino acids, but the mechanistic details of the reactions involved are unclear.

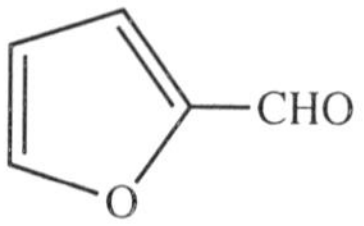

(1-12a)

$$\text{HOCH}_2-\underset{O}{\underset{|}{\text{furan}}}-\text{CHO}$$

(1-12b)

Intracellular reactions of carbohydrates are usually linked to metal ion-promoted oxidations involving iron and copper. Stabilized free radical intermediates or oxidized products of sugars, including various dicarbonyl intermediates, react with and damage proteins by crosslinking and condensation reactions (Wellsknecht et al., 1995).

Polymeric carbohydrates (starch, cellulose, hemicelluloses) play important roles both in living cells and in manufactured goods. The usual route of decay of cellulose-containing products, such as wood, paper, and fabrics, is hydrolysis, in which the monosaccharide or oligosaccharide units are cleaved from the polymer by (usually) acids. However, a number of studies of the autooxidative degradation of carbohydrate polymers have been performed. Exposure of cellulose to aerobic environments, for example, results in an increase in –COOH groups (Ott et al., 1954) as well as carbonyl (aldehyde and ketone) groups and aromatic rings (Whitmore and Bogaard, 1995; Norton et al., 1995).

In the presence of dyes, or of trace amounts of transition metal ions, oxidative degradation may be accelerated. These phenomena are clearly seen in examination of old documents where iron-containing inks have been used to print the text; the area in the vicinity of the ink is damaged, whereas the surrounding paper is more or less intact.

b. Vitamins. Studies of vitamin autooxidation are of two different types. A few vitamins, namely C and E, are well understood to be intracellular antioxidants. As readily oxidizable compounds, their *in vitro* and *in vivo* oxidative reactions have been thoroughly studied under a variety of conditions.

The decomposition of the other vitamins has been principally studied by food scientists and those interested in dietary supplements. These scientists have principally been concerned with monitoring the loss of vitamins from foodstuffs or tablets, and less engaged in studying the reactions leading to the loss of these compounds. Because foods are very complex environments, it is not surprising that the fates of many of these compounds are not understood in full molecular detail.

Folic acid (1-13) oxidatively degrades in the gastrointestinal tract, especially under the acid conditions typical of the stomach environment. The mechanism of the oxidation includes cleavage of the C9–N10 bond. Ascorbate is able to provide significant protection against decomposition (Lucock et al., 1995).

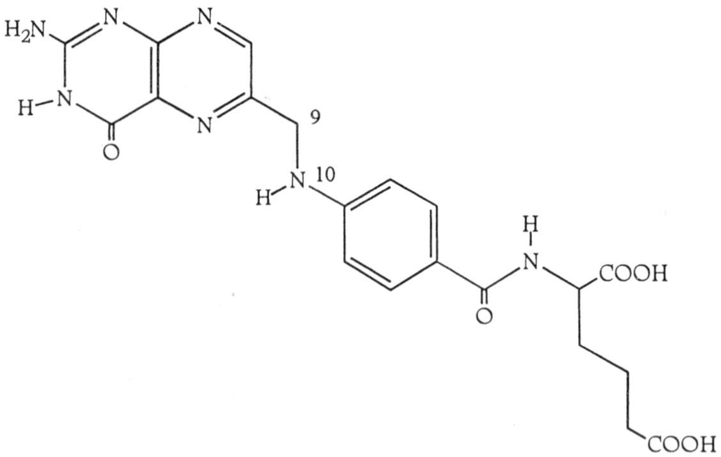

(1-13)

Vitamin D (1-14) is susceptible to photodecomposition, especially in the presence of photosensitizers, which bring about addition of an oxygen molecule across the exocyclic carbons of the A ring, leading to an endoperoxide (Yamada et al., 1978).

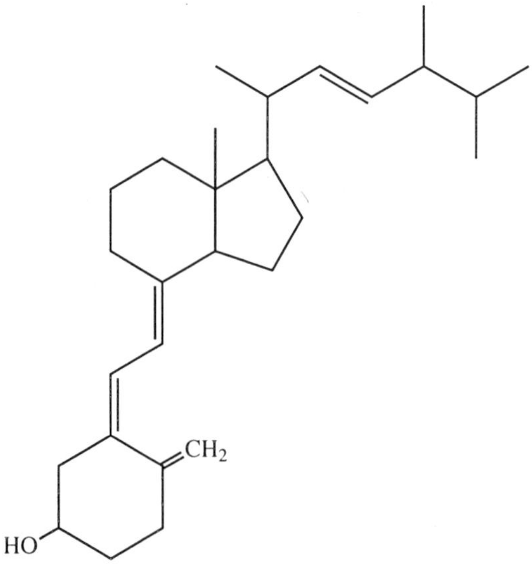

(1-14)

C. The Measurement of Autooxidation

A wide selection of methods for the analysis of biological and abiotic samples undergoing autooxidation is available. No single approach is applicable to all situations. Some techniques focus on measuring the loss of one of the reactants (oxygen or substrate) from the autooxidizing mixture, while others concentrate on characterizing or identifying products of the reaction. Among the latter are organoleptic tests, in which a panel of trained evaluators subjectively assesses the degradation of a material, such as a food product, by sensory evaluation criteria. Although sometimes useful, these methods will not be discussed further.

Evaluative methods may either be static or dynamic. In static methods, the autooxidized sample is tested at some single point in its degradative history and one or more properties of the sample is recorded. Dynamic methods are inherently kinetic, that is, a property that is changing as the sample oxidizes is monitored continuously. Although both approaches may be useful, dynamic techniques are more likely to give information of predictive value.

Many of these methods require expensive instrumentation, are susceptible to analytical errors, or otherwise are of limited utility. A full examination of the advantages and disadvantages of the major techniques is beyond the scope of this book, and only a brief discussion of some of the principal methods will be presented here. Fuller reviews are available (Gray, 1978: AOCS, 1989; Halliwell and Chirico, 1993; Aruoma, 1996).

1. Oxygen Uptake

Since autooxidation is accompanied by incorporation of atmospheric oxygen into the material, it is possible to follow the reaction by measuring the change of oxygen pressure in a sealed vessel. Various types of manometric devices have been used, including simple Warburg flask assemblies, microgas measuring burettes, and more elaborate assemblies that allow the user to manipulate the temperature, gas pressure, and light intensity (Grassie and Weir, 1965). Automatic measuring of the rate of oxygen uptake, using a pressure transducer attached to a recording device, is extremely useful in determining the kinetics of inhibited autooxidation. This property may also be measured using a special gas chromatographic reactor (Faria, 1982). Oxygen uptake can also be measured by recording the gain in weight of a substance (as long as volatile products are not being formed at the same time).

2. Peroxide Determination (Active Oxygen Method)

Introduction of oxygen into an oxidizing organic compound, at least in its initial stages, is usually accompanied by peroxide formation. A traditional method used widely by food scientists is the determination of peroxide value or PV. PV, expressed in milliequivalents of peroxide per kilogram of sample, is an officially sanctioned method for the analysis of oils and fats (AOCS,

1989). Peroxides may either be measured directly by spectroscopic, polarographic, or chromatographic methods, or, more typically, by their ability to bring about the oxidation of a substrate that can then be measured by some sensitive colorimetric or fluorimetric test.

Among the direct measurement tests are chromatographic measurements of particular groups of peroxidic species, usually hydroperoxides. Lipid hydroperoxides can be more or less specifically determined by HPLC (Ansari and Smith, 1979; Hughes et al. 1983; Funk and Baker, 1985) or by gas chromatography or GC-MS analysis of their derived alcohols (Hughes et al. 1986). Thin-layer chromatography, using spray reagents such as the Würster dyes that give intensely colored spots, is useful for rapid qualitative detection of oxidants (Smith and Hill, 1972).

Many substances that can be oxidized by peroxides, giving rise to colored or fluorescent products that can be quantitatively measured by titration or spectrometry. The official method for PV determination in foods relies on an iodimetric titration; the sample is treated with a saturated solution of potassium iodide and the liberated iodine is back-titrated with thiosulfate (AOCS, 1989). Iodine can also be detected colorimetrically (Siddiqui and Tappel, 1955; Takagi et al., 1978). In all methods that liberate iodine, an improvement in precision results when the analysis is conducted under conditions that prevent the undesirable oxidation of iodide by air oxygen, which can result in significant blank corrections.

Other spectroscopic and colorimetric assays have been reported that depend on the formation of strongly absorbing materials. One family of methods, which dates back many years, has exploited the ability of peroxides to oxidize ferrous (Fe^{2+}) iron salts to ferric (Fe^{3+}) species. It is generally recognized that these methods are highly sensitive relative to iodometry, but they can be subject to large errors, especially in aerated systems. The addition of specific Fe^{3+} complexing agents (such as thiocyanate, o-phenanthroline, etc.) has been reported to increase the sensitivity of these methods. One of the more interesting recent variations of such methods is the use of the dye xylenol orange as an indicator for the detection of oxidation products within lipoprotein samples (Jiang et al., 1992).

Some organic compounds also develop intense colors when oxidized. Polyphenols such as phloroglucinol (1-15), for example, form red products when attacked by peroxides. These compounds are measured spectrometrically in the classical Kreis test (Pool and Prater, 1945).

(1-15)

Fluorescence assays for peroxidic products have the potential for greatly increased sensitivity over other methods of analysis (Fletcher et al., 1973). One method of fluorescence determination of hydroperoxides is based on the oxidation of dichlorodihydrofluorescein to the corresponding fluorescein in the presence of hematin (Cathcart et al., 1983). Another fluorescence assay using a diphenyl pyrenyl phosphine, which when oxidized by lipid hydroperoxides gives rise to a highly fluorescent product, has also been described (Akasaka et al., 1992).

3. Formation of Malondialdehyde and Other TBA-Reactive Substances

For many years it has been recognized that during the course of lipid peroxidation, products are formed from polyunsaturated fatty acid precursors that include malondialdehyde, $OHC-CH_2-CHO$, and compounds related to it. Some of these compounds react with thiobarbituric acid (TBA) (1-16) to form colored products that can be measured by monitoring their absorption at around 530 nm (Gray, 1978). At least some of these compounds are Schiff bases formed by reactions such as those outlined in Equation 1-3.

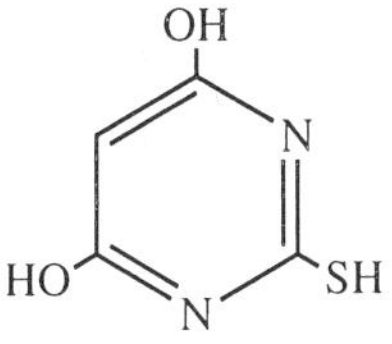

(1-16)

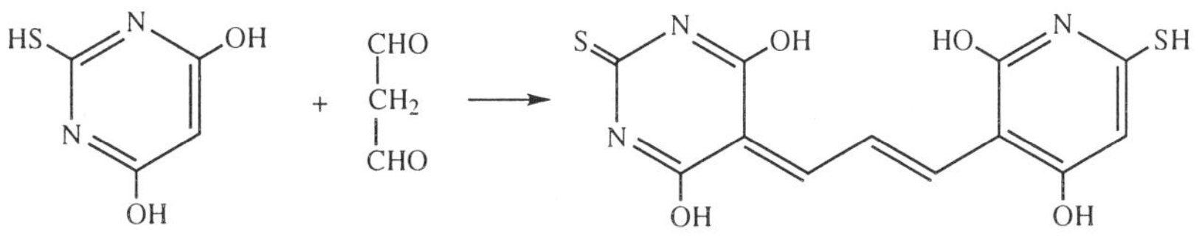

(Equation 1-3)

Because of the nonspecific nature of colored product formation, the data are usually reported as relative amounts of "TBARS" (TBA-reactive substances) formed under given exposure conditions. Alternatively, the "TBA number" of a sample is reported in terms of the color intensity afforded by a standard number of micrograms of malonaldehyde per gram of sample (or ppm of malonaldehyde). Only a small fraction of the products formed during

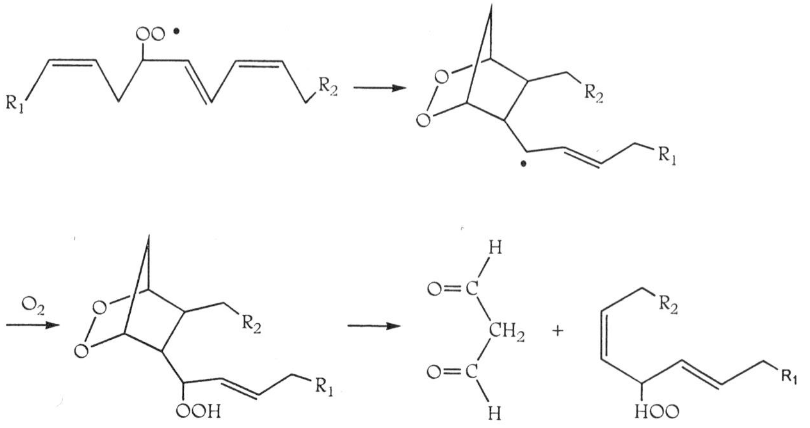

Figure 1-3 Mechanism for production of malondialdehyde from autooxidizing lipids as proposed by Pryor et al., 1976. (From Pryor, W. A., J. P. Stanley, and E. Blair, Lipids, 11, 370–379, 1976. With permission.)

lipid oxidation react with TBA. Nevertheless, because of the sensitivity and simplicity of the technique, it remains in wide use.

A mechanism for the formation of malondialdehyde from unsaturated precursors was proposed by Pryor et al. (1976) (Figure 1-3).

4. Detection of Other Carbonyl Compounds

The determination of aldehydes and ketones in autooxidized organic matter can be done using a variety of methods, which are, however, inherently not very specific. Colorimetic reagents like 2,4-dinitrophenylhydrazine, for example, give products with most kinds of carbonyl groups, but the spectroscopic characteristics of the derivatives (absorption maximum, molar extinction coefficient), as well as their stabilities in the mixture under test, may be quite variable. Aldehydes, especially α,β-unsaturated types, which are common constituents of oxidized lipids, react with anisidine (1-17) to form colored, Schiff-base products that absorb at around 350 nm (IUPAC, 1987).

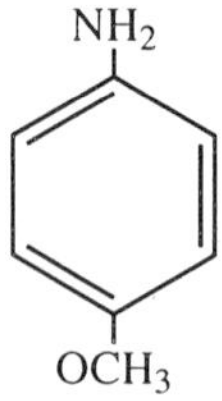

(1-17)

5. Conjugated Double Bond Formation in Lipids

Isolated double bonds, such as those of typical monounsaturated or 1,4-doubly unsaturated (allylic) fatty acids, do not absorb UV light above about 210 nm. Conversion of these compounds to free radicals, however, can result in extensive double bond migration into a more stable 1,3- (conjugated) configuration. There is a report, however, that a conjugated, but unoxidized, dienoic isomer of linoleic acid is a major constituent of human serum (Iversen et al., 1984). The absorption of these dienes at 233 nm has often been employed as a measurement of the extent of lipid peroxidation, especially in rather simple *in vitro* studies. However, the method is of limited usefulness in more complex materials such as foodstuffs or tissue extracts because of the presence of interfering chromophores (Hughes et al., 1983). A secondary method for conjugated double bonds makes use of the Diels-Alder reaction between a diene and a reactive acceptor, or dienophile, such as tetracyanoethylene (TCNE) (Waller and Recknagel, 1977) (Equation 1-4). In practice, the method relies on the detection of radiolabel derived from (^{14}C-TCNE)-lipid adducts. Because of various methodological difficulties and interference problems, this technique is seldom used.

(Equation 1-4)

More extensively autooxidized lipids also develop secondary chromophores, due to conjugated trienes and/or carbonyl groups, that begins to appear at about 265–285 nm. Although probably more diagnostic and less susceptible to interference from other biological chromophores, the intensity of this band is normally quite weak and therefore probably of limited value. Carbonyl values, however, are sometimes used as a rapid detection method either using direct UV spectrometry, as implied above, or by a colorimetric method involving derivatization (Lappin and Clark, 1951; Henick et al., 1954).

3. Aliphatic Hydrocarbon Production in Lipid Oxidation

One of the commoner methods for following the development of lipid peroxidation *in vivo* is the measurement of saturated hydrocarbons in the exhaled breath of the experimental animals (Dunnelin and Tappel, 1977; Müller and Sies, 1984), usually by gas chromatography. The compounds are also formed in plant tissues that have undergone oxidative damage (Peiser et al., 1982).

The generally accepted mechanism for the formation of these compounds is α-cleavage of a lipid-derived alkoxy radical (Tappel, 1980):

$$RCH_2-\overset{\overset{O\bullet}{|}}{CH}-CH_2R' \rightarrow RCH_2\bullet + OCHCH_2R' \xrightarrow{R^\wedge H} RCH_3$$

(Equation 1-5)

Alkanes are rather minor products of lipid oxidation in a quantitative sense, but because they are easily separable from the reaction mixture, their measurement is not too problematic. In most studies, ethane and pentane have been found to be the major alkanes expired. Several investigators, however, have pointed out that it is difficult to distinguish autooxidative damage in these sorts of studies from damage due to other deleterious cellular changes, and furthermore, that lipid damage could develop without necessarily leading to alkane production, a rather late stage in the generally accepted mechanistic pathway (Smith and Anderson, 1987).

7. Light Emission from Cells, Lipids, and Proteins

Biological tissues are known to contain fluorescent substances under normal conditions; however, some authors have claimed that unique light-emitting chromophores are generated during oxidative stress. Chemiluminescence has been attributed to singlet oxygen, excited carbonyl groups, and lipid peroxyl radicals (Cilento, 1982; Sies and Cadenas, 1985).

Detection of fluorescent products in biological tissues has the potential to be very sensitive; it has been claimed that the fluorescence method detects oxidation products at concentrations one or two orders of magnitude lower than the TBA test (Trombly and Tappel, 1975). However, the chemical nature of the fluorophores derived in *in vivo* experiments is still unclear.

8. Kinetic Methods

A family of quantitative methods has been developed for monitoring the rates of change in reference biomolecules undergoing autooxidation. These techniques, although perhaps more labor-intensive and mathematically sophisticated than many of the methods described above, are preferable in some ways because of the insights they give into the mechanisms of autooxidation and its inhibition by antioxidants. Analyzing the shapes of the kinetic curves, for example, tells something about the roles of free radical chains in the process, as well as the stoichiometries and rates of the agents taking part in the reactions. Details of the use and evaluation of these methods will be given in Chapter 4.

REFERENCES

Akasaka, K., I. Sasaki, H. Ohrui, and H. Meguro. 1992. A simple fluorometry of hydroperoxides in oils and foods. Biosci. Biotech. Biochem. 56, 605–607.

Anonymous. 1883. Autoxidation in living vegetable cells. Science 229, #2, 229–230.

Ansari, G. A. S. and L. L. Smith. 1979. High-performance liquid chromatography of cholesterol autoxidation products. J. Chromatogr. 175, 307–315.

AOCS (American Oil Chemists Society). 1989. Official methods and recommended practices. Am. Oil Chem. Soc., Champaign, IL.

Aruoma, O. I. 1996. Characterization of drugs as antioxidant prophylactics. Free Rad. Biol. Med. 20, 675–705.

Berlett, B.S., O. H. M. Omar, J. A. Sahakian, R. L. Levine, and E. R. Stadtman. 1991. Oxidation of proteins by ozone. Fed. Am. Soc. Exp. Biol. J. 5, A1524.

Bielski, B. H. J. and G. G. Shiue. 1979. Reaction rates of superoxide radicals with the essential amino acids. in Oxygen Free Radicals and Tissue Damage (CIBA Foundation symposium #65), Excerpta Medica, Amsterdam. pp. 43–56.

Blakely, W. F., A. F. Fulciarelli, B. J. Wegher, and M. Dizdaroglu. 1990. Hydrogen peroxide-induced base damage in deoxyribonucleic acid. Radiat. Res. 121, 338–343.

Buxton, G. V., C. L. Greenstock, W. P. Helman, and A. B. Ross. 1988. Critical review of rate constants for reactions of hydrated electrons, hydrogen atoms and hydroxyl radicals in aqueous solution. J. Phys. Chem. Ref. Data 17, 514–886.

Cathcart, R., E. Schwiers, and B. N. Ames. 1983. Detection of picomole levels of hydroperoxides using a fluorescent dichlorofluorescein assay. Anal. Biochem. 134, 111–116.

Cilento, G. 1982. Electronic excitation in dark biological processes. In Chemical and Biochemical Generation of Excited States, W. Adam and G. Cilento, eds. Academic Press, New York. pp. 277–307.

Davies, K. J. A. 1987. Protein damage and degradation by oxygen radicals. J. Biol. Chem. 262, 9895–9901.

Dean, R. T., J. V. Hunt, A. J. Grant, Y. Yamamoto, and E. Niki. 1991. Free radical damage to proteins: the influence of the relative localization of radical generation, antioxidants, and target proteins. Free Rad. Biol. Med. 11, 161–168.

de Saussure, T. 1820. Suite des observations sur les substances huileuses. Ann. Chim. Phys. 13, 337–361.

Devasagayam, T. P. A., S. Steenken, M. S. W. Obendorf, W. A. Schultz, and H. Sies. 1991. Formation of 8-hydroxy(deoxy)guanosine and generation of strand breaks at guanine residues in DNA by singlet oxygen. Biochemistry 30, 6283–6289.

Dumelin, E. E. and A. L. Tappel. 1977. Hydrocarbon gases produced during *in vitro* peroxidation of polyunsaturated fatty acids and decomposition of preformed hydroperoxides. Lipids 12, 894–900.

Faria, J. A. F. 1982. A gas chromatographic reactor to measure the effectiveness of antioxidants for polyunsaturated lipids. J. Am. Oil Chem. Soc. 59, 533–535.

Fletcher, B. L., C. J. Dillard, and A. L. Tappel. 1973. Measurement of fluorescent lipid peroxidation products in biological systems and tissues. Anal. Biochem. 52, 1–9.

Floyd, R. A., M. S. West, K. L. Eneff, and J. E. Schneider. 1989. Methylene blue plus light mediates 8-hydroxyguanine formation in DNA. Arch. Biochem. Biophys. 273, 106–111.

Fucci, L., C. N. Oliver, M. J. Coon, and E. R. Stadtman. 1983. Inactivation of key metabolic enzymes by mixed-function oxidation reactions. Possible implication in protein turnover and aging. Proc. Nat. Acad. Sci. USA 80, 1521–1525.

Funk, M. O. and W. J. Baker. 1985. Determination of organic peroxides by high performance liquid chromatography with electrochemical detection. J. Liquid Chrom. 8, 663–675.

Grassie, N. and N. A. Weir. 1965. The photooxidation of polymers. J. Appl. Polym. Sci. 9, 963–974.

Gray, J. I. 1978. Measurement of lipid oxidation: a review. J. Am. Oil Chem. Soc. 55, 539–546.

Gutter, B., W. T. Speck, and H. S. Rosenkrantz. 1977. Photodynamic modification of DNA by hematoporphyrin. Biochim. Biophys. Acta 475, 307–314.

Halliwell, B. and A. I. Aruoma. 1991. DNA damage by oxygen-derived species. Its mechanism and measurement in mammalian systems. FEBS Lett. 281, 9–19.

Halliwell, B. and S. Chirico. 1993. Lipid peroxidation: its mechanism, measurement, and significance. Am. J. Clin. Nutr. 57, 715S–722S.

Halliwell, B., J. M. C. Gutteridge, and C. E. Cross. 1992. Free radicals, antioxidants, and human disease: where are we now? J. Lab. Clin. Med. 119, 598–620.

Hazlewood, C. and M. J. Daviers. 1995. ESR spin-trapping studies of the reaction of radicals derived from hydroperoxide tumour-promoters with nucleic acids and their compounds. J. Chem. Soc. Perkin 2, 895–901.

Henick, A. S., M. F. Benca, and J. H. Mitchell. 1954. Estimating carbonyl compounds in rancid fats and foods. J. Am. Oil Chem. Soc. 31, 88–91.

Hodge, J. E. 1953. Dehydrated foods: chemistry of browning reactions in model systems. J. Agric. Food Chem. 1, 928–943.

Hoffman, A. W. 1860. Remarks on the changes of gutta percha under tropical influences. J. Chem. Soc. 13, 87–90.

Hughes, H., C. V. Smith, E. C. Horning, and J. R. Mitchell. 1983. HPLC and GC-MS determination of specific lipid peroxidation products *in vivo*. Anal. Biochem. 130, 431–436.

Hughes, H., C. V. Smith, J. O. Tsokos-Kuhn, and J. R. Mitchell. 1986. Quantification of lipid peroxidation products by gas chromatography-mass spectroscopy. Anal. Biochem. 162, 107–112.

IUPAC (Intl. Union of Pure Appl. Chem.). 1987. *Standard Methods for the Analysis of Oils, Fats, and Derivatives*. 7th ed. Oxford University Press, New York.

Iversen, S. A., P. Cawood, M. J. Madigan, A. M. Lawson, and T. L. Dormandy. 1984. Identification of a diene conjugated component of human lipid as octadeca-9,11-dienoic acid. FEBS Lett. 171, 320–324.

Jiang, Z. Y., J. V. Hunt, and S. P. Wolff. 1992. Ferrous ion oxidation in the presence of xylenol orange for the detection of lipid hydroperoxide in low density lipoprotein. Anal. Biochem. 202, 384–389.

Kanner, J., J. B. German, and J. E. Kinsella. 1987. Initiation of lipid peroxidation in biological systems. CRC Crit. Revs. Food Sci. Nutr. 25, 317–364.

Kappus, H. 1987. A survey of chemicals inducing lipid peroxidation in biological systems. Chem. Phys. Lipids 45, 105–115.

Lappin, G. R. and L. C. Clark. 1951. Colorimetric method for determination of carbonyl compounds. Anal. Chem. 23, 541–542.

Lee, P. C. C. and M. A. J. Rodgers. 1987. Laser flash photolysis kinetic studies of rose bengal sensitized photodynamic interactions of nucleotides and DNA. Photochem. Photobiol. 45, 79–86.

Lucock, M. D., M. Priestnall, I. Daskalakis, C. J. Shorah, J. Wild, and M. I. Levene. 1995. Nonenzymatic degradation and salvage of dietary folate: physicochemical factors likely to induce bioavailability. Biochem. Molec. Med. 55, 43–53.

Matheson, I. B. C. and J. Lee. 1979. Chemical reaction rates of amino acids with singlet oxygen. Photochem. Photobiol. 29, 879–881.

McCann, H. G. 1978. *Chemistry Transformed: The Paradigmatic Shift from Phlogiston to Oxygen.* Ablex Pub. Co., Norwood, NJ.

Mouret, J.-F., M. Polverelli, F. Sarrazini, and J. Cadet. 1991. Ionic and radical oxidations of DNA by hydrogen peroxide. Chem.-Biol. Interact. 77, 187–201.

Müller, A. and Sies, H. 1984. Assay of ethane and pentane from isolated organs and cells. Meth. Enzymol. 105, 311–319.

Namiki, M. 1988. Chemistry of Maillard reactions: recent studies of the browning reaction and the development of antioxidants and mutagens. Adv. Food Res. 32, 115–184.

Neužil, J., J. M. Gebicki, and R. Stocker. 1993. Radical-induced chain oxidation of proteins and its inhibition by chain-breaking antioxidants. Biochem. J. 293, 601–606.

Norton, F. J., G. D. Love, A. J. McKinnon, and P. J. Hall. 1995. Mechanism of char formation from oxidized cellulose. J. Mater. Sci. 30, 596–600.

Ott, E., H. M. Spurlin, and M. W. Graffin. 1954. *Cellulose and Cellulose Derivatives.* Interscience, New York.

Payne, J. R. and C. R. Phillips. 1985. Photochemistry of petroleum in water. Environ. Sci. Technol. 19, 569–579.

Peiser, G. D., M. C. C. Lizada, and S. F. Yang. 1982. Sulfite-induced lipid peroxidation in chloroplasts as determined by ethane production. Plant Physiol. 70, 994–998.

Piette, J. 1991. Biological consequences associated with DNA oxidation modified by singlet oxygen. J. Photochem. Photobiol. B11, 241–260.

Pool, M. F. and A. N. Prater. 1945. A modified Kreis test suitable for photocolorimetry. Oil and Soap 22, 215–216.

Pryor, W. A., J. P. Stanley, and E. Blair. 1976. Autoxidation of polyunsaturated fatty acids: II. A suggested mechanism for the formation of TBA-reactive materials from prostaglandin-like endoperoxides. Lipids 11, 370–379.

Razskazovskii, Y. and M. D. Sevilla. 1996. Reactions of sulphonyl peroxyl radicals with DNA and its components. Hydrogen abstraction from the sugar backbone versus addition to pyrimidine double bonds, Int. J. Radiat. Biol. 69, 75–87.

Rhaese, H. J. 1968. Chemical analysis of DNA alterations. III. Isolation and characterization of adenine oxidation products obtained from oligo- and monodeoxyadenylic acids treated with hydroxyl radicals. Biochim. Biophys. Acta 166, 311–326.

Schweitz, H. 1969. Dégradation du DNA par H_2O_2 en présence d'ions Cu^{++}, Fe^{++} et Fe^{+++}. Biopolymers 8, 110–119.

Shi, X. L., H. G. Jiang, Y. Mao, J. P. Ye, and U. Saffiotti. 1996. Vanadium(IV)-mediated free radical generation and related 2′-deoxyguanosine hydroxylation and DNA damage. Toxicology 106, 27–38.

Siddiqui, A. M. and A. L. Tappel. 1955. Colorimetric determination of lipid and organic peroxides. Chemist Analyst 44, 52.

Sies, H. and E. Cadenas. 1985, Oxidative stress: damage to intact cells and organs. Phil. Trans. R. Soc. Lond. B311, 617–631.

Simic, M. G. 1988. Mechanisms of inhibition of free-radical processes in mutagenesis and carcinogenesis. Mutat. Res. 202, 377–386.

Shigenaga, M. K. and B. N. Ames. 1991. Assays for 8-hydroxy-2′-deoxyguanosine: a biomarker of *in vivo* oxidative DNA damage. Free Rad. Biol. Med. 10, 211–216.

Smith, C. V. and R. E. Anderson. 1987. Methods for determination of lipid peroxidation in biological samples. Free Rad. Biol. Med. 3, 341–344.

Smith, L. L. and F. L. Hill. 1972. Detection of sterol hydroperoxides on thin-layer chromatoplates by means of the Würster dyes. J. Chromatogr. 66, 101–109.

Sohal, R. S., S. Agarwal, A. Dubey, and W. C. Orr. 1993. Protein oxidative damage is associated with life expectancy of houseflies. Proc. Nat. Acad. Sci. USA 90, 7255–7259.

Stadtman, E. R. 1992. Protein oxidation and aging. Science 257, 1220–1224.

Takagi, T., Y. Mitsuno, and M. Masumura. 1978. Determination of peroxide value by the colorimetric iodine method with protection of iodide as cadmium complex. Lipids 13, 147–151.

Tappel, A. L. 1980. Measurements of and protection from lipid peroxidation. In *Free Radicals in Biology*, Vol. 4, W. A. Pryor, ed. Academic Press, New York. pp. 2–48.

Trombly, R. and A. L. Tappel. 1975. Fractionation and analysis of fluorescent products of lipid peroxidation. Lipids 10, 441–447.

von Sonntag, C. 1987. *The Chemical Basis of Radiation Biology.* Taylor and Francis, London.

Waller, R. L. and R. O. Recknagel. 1977. Determination of lipid conjugated dienes with tetracyanoethylene-^{14}C: significance for study of the pathology of lipid peroxidation. Lipids 12, 914–921.

Ward, J. F. 1981. Some biochemical consequences of the spatial distribution of ionizing radiation-produced free radicals. Radiat. Res. 86, 185–195.

Wellsknecht, K. J., D. V. Zyzak, J. E. Litchfield, S. R. Thorpe, and J. W. Baynes. 1995. Mechanism of autoxidative glycosylation: identification of glyoxal and arabinose as intermediates in the autoxidative modification of proteins by glucose. Biochemistry 34, 3702–3709.

Whitmore, P. M. and J. Bogaard. 1995. The effect of oxidation on the subsequent oven aging of filter paper. Restaurator 16, 10–30.

Wolff, S. P. and R. T. Dean. 1987. Glucose oxidation and protein modification. The potential role of autoxidative glycosylation in diabetes. Biochem. J. 245, 243–250.

Wolff, S. P., M. J. C. Crabbe, and P. J. Thornalley. 1984. The autoxidation of glyceraldehyde and other simple monosaccharides. Experientia 40, 244–246.

Wolff, S. P., A. Garner, and R. T. Dean. 1986. Free radicals, lipids, and protein breakdown. Trends Biochem. Sci. 11, 27–31.

Yamada, S., K. Nakayama, and H. Takayama. 1978. Synthesis of 6,19-epidioxy-9,10-secoergosta-5(10), 7,22-trien-3β-ols from vitamin D derivatives by oxidation with singlet oxygen. Tetrahedron Lett. 49, 4895–4898.

Zirlin, A. and M. Karel. 1969. Oxidation effects in a freeze-dried gelatin-methyl linoleate system. J. Food Sci. 34, 160–164.

2 AUTOOXIDATION MECHANISMS

I. FREE RADICAL CHAIN REACTIONS

A. Nature of Radical Species

The electrons of most stable molecules are paired; that is, they occupy orbitals that contain two electrons having opposite spins. These orbitals may be combined to form covalent bonds (usually having two, four, or six electrons occupying the space between two atoms), or they may be in nonbonded pairs, such as those found on the oxygen atom of water. However, there are other chemical species in which one (or, uncommonly, more than one) electron is unpaired, or alone in its orbital. These species are referred to as "radicals"* or "free radicals."

Usually, radicals are designated using a single-dot convention to indicate the unpaired electron. Common examples of simple free radicals include the chlorine atom, $Cl\cdot$; the hydroxyl radical, $HO\cdot$; the superoxide anion, $\cdot O_2^-$; and the ground-state oxygen molecule, $\cdot O\text{–}O\cdot$ (the latter being technically a diradical, since an unpaired electron is localized on each of two separate atoms).

A free radical is produced whenever a covalent single bond between two atoms is cleaved in such a way as to leave at least one electron in an unpaired state. This process is referred to as homolysis if the bond breaks in such a way as to leave two unpaired partners. Normally, homolysis requires a fairly substantial input of energy, or "bond dissociation energy," which must be at least equal to, and usually is significantly greater than, the bond energy of the bond in question. A typical homolysis reaction is the cleavage, by heat or UV light, of the O–O bond in peroxides such as H_2O_2:

---

* Historically in chemistry, the name **radical** was applied to a group of atoms that remained unchanged during a reaction. In organic chemistry, radicals were assumed to be bound, in the sense that a hydroxyl group, for example, was covalently attached to a carbon chain. When Gomberg, in 1900, suggested that organic radicals might be capable of independent existence, or "free," his suggestion was not widely accepted for many years. Subsequently, the former usage of "radical" died out. We now use the term "group" or "functional group" for this sense; "radical" is now used exclusively to refer to odd-electron species.

$$HO-OH \xrightarrow{h\upsilon} 2\ HO\cdot$$

(Equation 2-1)

An O–O bond typically has a dissociation energy of about 200 kJ/mol, and its homolysis can proceed at relatively low temperatures or with absorbed light of relatively long wavelengths. Dissociations of stronger bonds, such as C–C or C–H, certainly can occur, but require considerably higher inputs of energy. To cleave a typical C–H bond requires an input of about 400 kJ/mol, which is equivalent to a temperature of around 550°C. Temperatures of this magnitude are encountered in combustion reactions, or in industrial processes like the thermal cracking of hydrocarbons.

Another way of producing a free radical is for a pre-formed odd-electron species to react with an even-electron compound. For example, a hydroxyl radical may add to a carbon-carbon double bond. A new covalent bond is formed between the HO· and one of the double bond electrons; the other electron becomes "free:"

$$HO\cdot\ +\ >C=C<\ \rightarrow\ HO-\overset{|}{\underset{|}{C}}-\overset{|}{\underset{|}{C}}\cdot$$

(Equation 2-2a)

Another important process, especially in biological systems, is the addition of an electron (the ultimate odd-electron species!) to a multiple bond. In this process the acceptor molecule becomes negatively charged. Quinones, for example, may be reduced by successive one-electron transfers (Equation 2-2b):

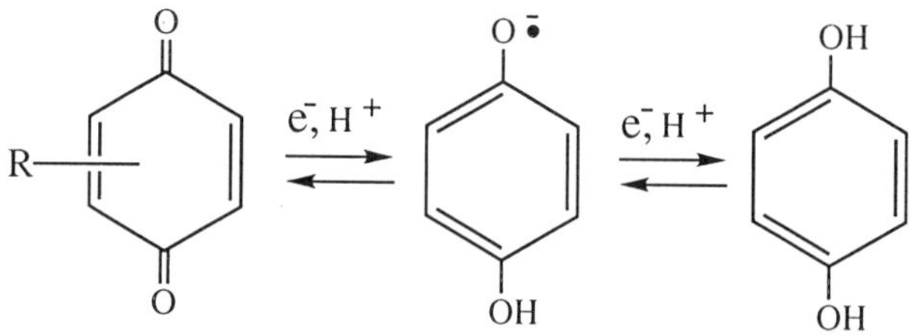

(Equation 2-2b)

The intermediate radical anion, which often is somewhat stable, has the trivial name *semiquinone* in this case. Biological quinones and related compounds are often reduced by reductase enzymes that work by means of this mechanism. Note that the one-electron oxidation of a hydroquinone would also produce a semiquinone radical.

The above reactions are of the type A· + B → AB·, where A is the odd-electron species and B the even-electron form. Or, in a slightly different process, the odd-electron species can remove a group from, rather than add to, molecule B. This might be generalized as A· + BX → AB· + X⁻. These reactions are analogous to nucleophilic displacements. An example is the addition of an electron to CCl_4:

$$e^- + CCl_4 \rightarrow \cdot CCl_3 + Cl^-$$

(Equation 2-3)

Transition metals whose valence states differ by one unit also can act as radical precursors, as illustrated by the traditional mechanism for the redox reaction of iron(II) and H_2O_2.

$$Fe(II) + HO\text{-}OH \rightarrow Fe(III) + HO^- + HO\cdot$$

(Equation 2-4)

Still another mode of radical generation is the "unpairing" of electrons in multiple bonds. A carbonyl ($R_2C=O$) group, for example, absorbs ultraviolet light of certain wavelengths and is converted to an excited state. In quantum-mechanical terms, this is described as a conversion of a bonding to an anti-bonding orbital. A carbonyl group photoexcited in this fashion has some of the characteristics of a diradical, with a structure that can be approximated as $R_2C\cdot\text{-}O\cdot$; such a radical, with adjacent odd-electron atoms, is very reactive, and can participate in reactions such as the abstraction of a hydrogen atom from a suitable donor compound:

$$R_2C\cdot\text{-}O\cdot + R'H \rightarrow R_2C\cdot\text{-}OH + R'\cdot$$

(Equation 2-5)

Note that in this case the diradical gives rise to two new monoradicals.

Radicals may also be formed by the ejection of a single electron from a molecule, leaving behind a residual, positively charged odd-electron species referred to as a radical cation. Such processes are common in radiation chemistry and in mass spectrometry, where molecules are bombarded with high-energy particles. However, they can also occur in more environmentally and biologically relevant environments. Photoionization, for example, is a common fate of electron-rich compounds. The ejection of an electron from an excited state results in the formation of a radical cation; the ejected electron can either be taken up by the solvent or by another acceptor molecule. This is illustrated below for a phenol.

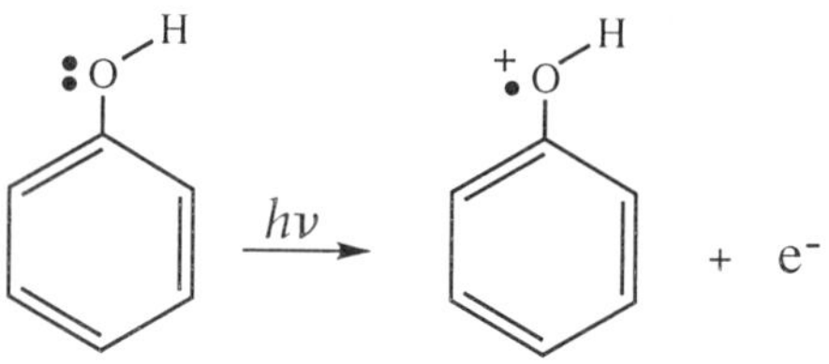

(Equation 2-6)

Water is particularly efficient in stabilizing electrons, and the "hydrated electron," or e_{aq}, is a well-known species. This substance was first produced in ionizing radiation experiments. Water, bombarded with (for example) a pulse of gamma rays, may lose an electron, leaving behind its radical cation:

$$H_2O \rightarrow H_2O^+ + e^-$$

(Equation 2-7)

The ejected electron is stabilized, or hydrated, by surrounding water molecules that help keep it away from cationic species. It is designated e_{aq} to convey this hydrated configuration.

Oxygen is a very reactive acceptor of electrons or hydrated electrons, with the product being $\cdot O_2^-$ (or superoxide anion, which, like a semiquinone, is an example of a radical anion). These reactions can either occur unimolecularly (that is, the donor of the electron absorbs light and emits the electron) or bimolecularly, where the donor compound and acceptor oxygen form a charge transfer complex which is the light-absorbing species, and electron transfer occurs within the complex (Mattay, 1987). Electron transfer to substrates other than oxygen can also be greatly enhanced by preliminary complex formation (Mattes and Farid, 1982). If the partner species separate, two new free radicals (a radical anion and cation) are released into solution (Eriksen et al., 1977).

Free radicals are usually quite reactive species, although some are stabilized to the point where they can be stored for years without reaction. Certainly ordinary molecular oxygen, $\cdot O-O \cdot$, is a very stable diradical. In addition, organic free radicals such as diphenylpicrylhydrazyl (structure 2-1) are capable

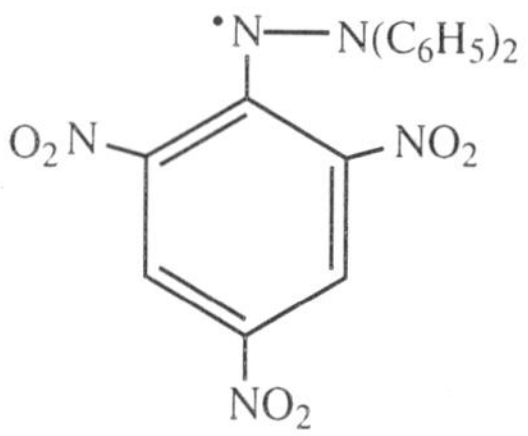

(2-1)

of prolonged existence. This large increase in stability may be due either to steric factors (the radical is so hindered as to be unable to collide in a chemically productive way with surrounding reagents) or to electronic effects, such as delocalization of the radical density over a large number of atoms. Also, in large organic polymers such as melanin, soot, and humic materials, stable radicals may continually be present at a low but measurable level.

B. Formation of Free Radicals (Initiation)

Investigations of the oxidative decomposition of both natural and synthetic materials have clearly shown that they proceed by free-radical processes. Radicals are present throughout the history of any complex substance; they are constantly being produced within the material by numerous routes, many of which have already been mentioned. In addition, they are continually being destroyed by chemical reactions. The relative rates of these production and destruction reactions govern whether a substance remains more or less intact and useful, or whether it is damaged.

Single-electron or odd-electron processes are in contrast to the more familiar electron-pair transfers that take place in, for example, many electrophilic and nucleophilic substitutions. Nevertheless, the understanding of these reactions is not fundamentally more difficult as long as one keeps in mind the truism "odd + even = odd", which, in chemical terms, states that a reaction between an odd-electron species (free radical) and even-electron species always results in the formation of a new odd-electron product. (Only when two odd-electron forms interact does a free radical reaction terminate. Note that this is simply the reverse of a reaction that forms radicals.)

It is possible, under favorable conditions, for odd-electron species to give rise to other such species in a series of repeating reactions that may go on for thousands or even millions of individual steps. This type of reaction is called a free radical chain reaction. Examples are seen in industrial processes involving radical initiators, such as the synthesis of polyethylene from ethylene, and also in environmental and biological processes, such as the breakdown of rubber or the peroxidation of lipids.

One group of initiators that is very widely used consists of azo compounds, that is, substances containing a central $-N=N-$ linkage. When heated to moderate temperatures, they eliminate molecular N_2 and produce two moles of free radicals:

$$R-N=N-R \rightarrow 2\,R\cdot\, +\, N\equiv N$$

(Equation 2-8)

Further details of the chemistry of these compounds are given in Chapter 4. For our current purposes, it will be sufficient to say that radicals produced during the decomposition of azo initiators add to monomeric compounds having free vinyl ($CH_2=$) groups to give new radicals. The derived radicals continue to add monomer units until very long chain polymers are produced:

$$R\cdot + CH_2 = CHX \rightarrow R-CH_2-\dot{C}HX \xrightarrow{CH_2=CHX} R-CH_2-CHX-CH_2-CHX\cdot \rightarrow \text{etc.}$$

(X may be $-H$, $-CH_3$, $-Cl$, -phenyl, etc.)

(Equation 2-9)

Note that each step of the sequence is an example of the $A\cdot + B \rightarrow AB\cdot$ or "odd + even = odd" reaction type. The kinetic chain in this type of polymer-forming reaction is related to the buildup of the actual chain, or polymer, that is formed.

Another type of important chain reaction is the nonpolymer-forming autooxidation of aliphatic hydrocarbons, RH (which do not contain double bonds or functional groups). This type of reaction may have a kinetically long chain, but no large molecules are formed; in fact, a reverse process resulting in the degradation of the starting material occurs. The chain events occur because radicals are transferred to new molecules rather than being retained in the same molecule.

This type of chain reaction is normally inaugurated in the liquid phase by a chemically reactive initiating agent. The initiator for these reactions is ordinarily a reactive free-radical oxidant capable of abstracting a hydrogen atom (or an electron) from the starting material. It may be a material that is already present in the environment of the autooxidizable compound, such as a Cu(II) atom; or one generated thermally, radiolytically, or photochemically in the environment of the substrate by some reaction; or it may be a pre-existing free radical that migrates to the reaction site, such as $\cdot OH$. In any event, the reaction

$$In + RH \rightarrow R\cdot + InH$$

(Equation 2-10)

(where In = the initiator species) produces a new free radical, $R\cdot$, that is often capable of taking part in further reactions leading to oxidized products.

In the initiated autooxidation of hydrocarbons and other substrates, the first (induction or initiation) phase of the reaction is typically quite slow, as shown in Figure 2-1.

The details of this kinetic curve are described more fully in later sections of this chapter and in the next chapter.

The slowness of this first step in the process is a consequence of a low steady-state concentration of free radicals in the beginning of the reaction. In Equation 2-10, it can be seen that the initiator, which usually is itself a radical, merely forms a new odd-electron species from the substrate, and is itself

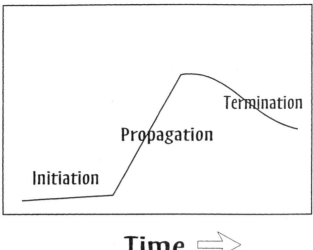

Figure 2-1 Typical course of a free-radical chain reaction. Vertical axis represents the magnitude of a property such as oxygen uptake that occurs concomitantly with substrate oxidation.

converted to a much less reactive, even-electron species. In order for the rate of the chain reaction to accelerate, there must be a mechanism for increasing the concentration of reactive free radicals.

C. Reactions of Initiator Free Radicals with Substrates (Propagation)

Although some free radicals are stable, most are rather reactive. In particular, hydrocarbon-derived radicals R· may combine with one another by simple covalent bond formation (odd + odd = even) to dimers R–R. For example, methyl radicals may dimerize to form ethane. Other possible reactions in which alkyl radicals may possibly engage include reactions with quinones, elemental sulfur, aromatic amines, and a few other compounds (Franz et al., 1987: Burton *et al.*, 1996: Neta *et al.*, 1996).

Usually, however, sufficient oxygen is present to permit the very fast reaction

$$R· + O_2 \rightarrow ROO·$$

(Equation 2-11)

to predominate. The rate of Reaction 2-11 is many orders of magnitude faster than dimerization. (This makes oxygen an important inhibitor of the radical chain polymerization of olefins such as styrene. However, the same property is favorable for oxidative decomposition pathways.) The reason is that molecular oxygen is itself a diradical (having the approximate structure ·O–O·) and therefore does not need to overcome a spin barrier in its reactions with other radicals. The kinetics of the reaction, which often approaches the theoretical maximum limit, is discussed in Chapter 4.

The intermediate ROO· is called a peroxyl radical; it is a key species in many biological and abiotic oxidation processes. Although an oxidizing radical, it is much more selective than HO· and, for example, tends to attack tertiary carbon-hydrogen bonds in preference to secondary or primary types; and even these reactions are very slow.

Peroxyl radicals have several fates, but the reaction that is most important for the majority of metabolic and non-biological oxidation reactions is a hydrogen abstraction:

$$ROO· + R'H \rightarrow ROOH + R'·$$

(Equation 2-12)

The product, ROOH, is called a hydroperoxide. The rates of these reaction types depend on temperature and on the structures of the reactants, but are usually far slower than that of the ROO·-forming reaction (Equation 2-11). It is favored in situations where hydrogen donors are abundant, for example within living cells or in polymers. Reaction 2-12 is called a chain-carrying reaction, because it generates a new species, R'·, that is capable of subsequent reaction with molecular oxygen to generate more ROOH.

In addition, hydroperoxides themselves are subject to decomposition to give additional free radicals:

$$ROOH \rightarrow RO· \text{ and/or } ·OH$$

(Equation 2-13)

The O–O bond in most hydroperoxides is quite weak (~130 kJ/mol; compare that in H_2O_2, about 220 kJ/mol) and therefore susceptible to cleavage. The fragmentation can be promoted by heat, metal ion catalysis, or UV light absorption.

There is also a bimolecular decomposition:

$$2\ ROOH \rightarrow H_2O + RO· + ROO·$$

(Equation 2-14)

This reaction appears to be more thermodynamically favored than the unimolecular decomposition, but requires higher ROOH concentrations than are usually encountered to become important.

Formation of two new odd-electron forms from the even-electron compound, ROOH, is a critical step in speeding the rate of hydrocarbon oxidation.

AUTOOXIDATION MECHANISMS

As hydroperoxides decompose, their radical products increase the concentration of chain-carrying radicals in the reaction, accelerating the rate of chain initiation, and therefore accelerating the overall rate of autooxidation.

Metal catalysis is often critically important because, depending on the metal ion, either a peroxyl or a more reactive alkoxy radical may be produced:

$$ROOH + M^{n+1} \rightarrow ROO\cdot + M^{n+} + H^+$$

(Equation 2-15a)

$$ROOH + M^{n+} \rightarrow RO\cdot + M^{n+1} + HO^-$$

(Equation 2-15b)

Metal ions that are strong oxidants, such as Pb^{4+}, tend to react predominantly by Reaction 2-15a, whereas in the case of strongly reducing species like Fe^{2+}, Reaction 2-15b predominates. (Note that Reaction 2-15b is identical to the traditional Fenton reaction, Reaction 2-40, if R=H.) The two reactions may in favorable cases (when the metal ion has valence states of similar stability, such as Co^{2+}/Co^{3+}) be summed, leading to an autocatalytic decomposition of hydroperoxide into both kinds of radical (Reich and Stivala, 1969).

The products of decomposition of the relatively unreactive hydroperoxide by these mechanisms include one or more equivalents of highly reactive radicals, such as ROO·, RO· (alkoxy radical) or ·OH. Therefore, a net increase in the yield of free radical species from the precursor, RH, occurs. This process, known as chain branching, is largely responsible for the rate acceleration observed in the autocatalytic stage in the traditional kinetic curve for free radical-induced oxidations (Figure 2-1).

Peroxyl radicals are also capable of reaction with olefins, producing intermediate β-alkylperoxyl radicals:

$$ROO\cdot + >C=C< \rightarrow ROO-\overset{|}{\underset{|}{C}}-\overset{\bullet}{C}<$$

(Equation 2-16)

These radicals, in aerobic environments, normally react with another molecule of molecular oxygen, but at lower O_2 concentrations epoxides may form:

$$ROO-\overset{|}{\underset{|}{C}}-\overset{\bullet}{C}< \rightarrow RO\cdot + >\overset{O}{\overset{/\,\backslash}{C-C}}<$$

(Equation 2-17)

The relative rates of these reactions are very dependent on the structure of the olefin and on its surroundings (see Chapter 4).

D. Destruction of Free Radicals (Termination)

A radical chain reaction ends, or becomes immeasurably slow, either when the concentration of odd-electron species falls to zero, or when no more chain-carrying species can be formed. Usually, the latter process is the governing step. In a mixture of hydrocarbons, for example autooxidizing petroleum, those hydrogen atoms that are most readily abstracted are kinetically selected early in the sequence. The less reactive hydrogens that remain are then not able to compete with radical-destroying reactions between the odd-electron species. Therefore, in the final phase of an initiated autooxidation, the availability of R'H (Reaction 2-12) decreases, and radicals tend to survive long enough to react with one another. Since radicals have an odd number of electrons, and are usually not charged, there is neither a coulombic nor a spin barrier to reactions between them, resulting in activation energies that are small or nonexistent, and therefore reactions that are kinetically fast. Often, termination rates are governed not by rates of reactions between radicals, but by mass transfer parameters such as the viscosity of the solvent. When collisions occur, two equivalents of odd-electron species are destroyed (odd + odd = even), and the concentration of radicals in the reaction mixture declines to a concentration that becomes insignificant.

Odd-electron combination reactions are of several types. Most simply, two atoms having unpaired electrons can simply add to one another to form a new compound with a covalent bond:

$$2 \cdot CH_3 \rightarrow H_3C-CH_3$$

(Equation 2-18)

A second type of reaction is disproportionation, or hydrogen atom transfer between one radical and another. Two moles of even-electron forms are produced from two moles of odd-electron species. This is a self-redox reaction:

$$2\ CH_3CH_2 \cdot \rightarrow H_2C=CH_2 + CH_3CH_3$$

(Equation 2-19)

or

$$2\ semiQ \rightarrow Q + HQ$$

(Equation 2-20)

Finally, the radical may transfer an electron to an oxidizing agent or accept one from a reducing agent:

$$ROO \cdot + Fe^{2+} + H+ \rightarrow ROOH + Fe^{3+}$$

(Equation 2-21)

The rates of recombination of various radical types are highly variable and dependent to a great extent on their structures. In general, the recombination rates between simple alkyl radicals, R·, are very high, but these radicals are almost never important in chain reactions unless molecular oxygen is absent.

The reactions of peroxyl radicals, because of their relatively high steady-state concentrations and moderate reactivities, tend to predominate in the termination phase. Some of the principal reaction types of peroxyl radicals have been summarized by von Sonntag (1987) and by Rånby and Rabek (1975). Considerable data has been accumulated on the rate constants of peroxyl radical reactions (Hendry et al., 1974: Willson, 1985; Neta et al., 1990). Because of interest in chloroform and carbon tetrachloride toxicity, rate data for the trichloromethylperoxyl radical ($Cl_3COO\cdot$) are common (Table 2-1), especially for reactions with biochemically important substrates. However, this radical is typically between one and two orders of magnitude more reactive than the usual peroxyl radical (Simic and Hunter, 1983).

Table 2-1 Approximate Second-Order Rate Constants for the Reaction of $Cl_3COO\cdot$ Radical with Various Substrates

Compound	Rate constant
Ascorbic acid	2×10^8
β-Carotene	1.5×10^9
Indole	1×10^8
Linoleic acid	1×10^6
Linolenic acid	4×10^6
Methionine	3×10^7
Oleic acid	1×10^6
Phenol	$<1 \times 10^5$
Pyridine	7×10^3
Tryptophan	9×10^7
Uric acid	7×10^8
Vitamin E	3.5×10^8

Note: Units, L mol^{-1} sec^{-1}.

Sources: Willson, 1985; Neta et al., 1990.

If the concentration of oxygen is sufficient, the principal radical-radical termination reaction is:

$$2\ ROO\cdot \rightarrow [ROO\text{–}OOR] \rightarrow products$$

(Equation 2-22)

The intermediate tetroxide is often referred to as the Russell tetroxide, after the free-radical chemist who first proposed its existence (Russell, 1957). The reaction as shown is reversible in some cases, especially for hindered secondary or tertiary alkylperoxyl radicals (Furimsky et al., 1980). Tetroxides can also decompose by at least three pathways, depending on their structures. First, they may break down heterolytically, with the loss of molecular oxygen, to afford two alkoxy radicals:

$$ROO\text{–}OOR \rightarrow 2\ RO\cdot + O_2$$

(Equation 2-23)

If α-hydrogen atoms are available, concerted processes can be envisaged to illustrate the conversion of the tetroxide to H_2O_2 and two equivalents of carbonyl compounds (Equation 2-24), or, alternatively, to oxygen and one equivalent each of a carbonyl compound and alcohol (Equation 2-25):

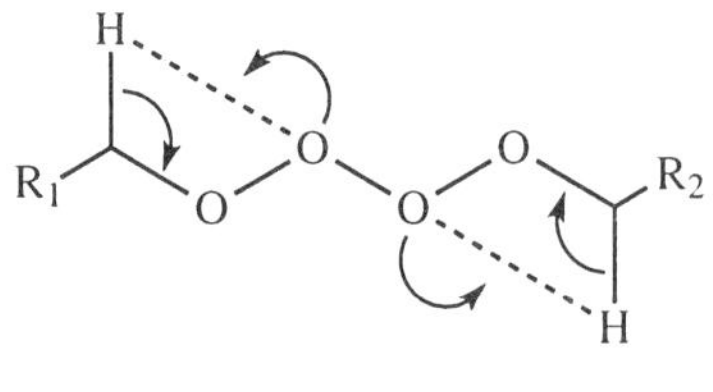

(Equation 2-24)

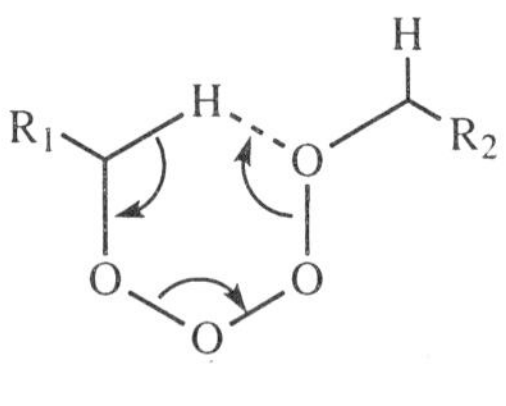

(Equation 2-25)

Some peroxyl radicals, such as α-hydroxy peroxyl radicals, decompose with the liberation of HOO·, which disproportionates to form H_2O_2 and O_2:

$$R_2\overset{\overset{OH}{|}}{C}-OO\cdot \rightarrow R_2C=O + HOO\cdot$$

(Equation 2-26)

A similar reaction occurs with β-hydroxyperoxyl radicals, such as those that are derived from subsequent additions of HO· and O_2 to a benzene ring:

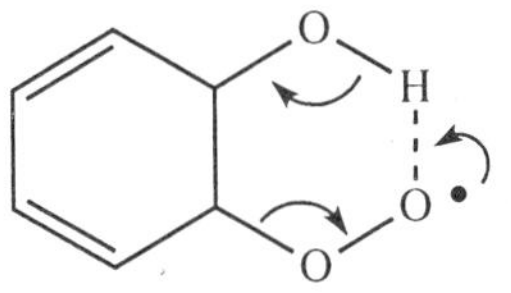

(Equation 2-27)

At low oxygen concentrations, peroxyl radicals may also react with alkyl radicals to form diperoxides:

$$ROO\cdot + R\cdot \rightarrow ROOR$$

(Equation 2-28)

Alkoxy radicals may react with one another by disproportionation. The electron transfer induces one equivalent of oxidized product (carbonyl) and one equivalent of reduced product (alcohol) to form:

$$2\ RO\cdot \rightarrow R=O + ROH$$

(Equation 2-29)

In certain cases, favorable elimination reactions involving alkoxy radicals may be important. For example, homolysis of a C–C bond β to the radical may take place, with the elimination of a carbonyl compound such as formaldehyde and the formation of a new carbon-centered radical (Nieman, 1964):

$$R_2CH-CH_2O\cdot \rightarrow R_2CH\cdot + CH_2=O$$

(Equation 2-30)

Another possibility is the dimerization of alkoxy radicals to peroxides,

$$2\ RO\cdot \rightarrow ROOR$$

(Equation 2-31)

or, hydroxyl and tertiary alkoxyl radicals may recombine:

$$HO\cdot + HCR_2-O\cdot \rightarrow H_2O + R_2C=O$$

(Equation 2-32)

The latter two reactions are usually restricted to cases in which the hydroxyl and alkoxyl radicals are generated in close proximity to one another, such as when a tertiary hydroperoxide or Russell tetroxide decomposes homolytically.

Some metal ions, for example Cu^{2+} and Co^{3+}, when present in high enough concentrations, may act as inhibitors as well as promoters of autooxidation. This is probably an example of electron transfer from or to $ROO\cdot$, as described earlier. Several investigators have noted that a "critical concentration" of metal catalyst is sometimes observed, above which a very steep drop in the rate of oxidation occurs. The explanation for this effect is not completely elucidated, but some researchers have postulated the formation of complexes in which $ROO\cdot$ becomes a ligand for the metal (Reich and Stivala, 1969).

II. SINGLET OXYGEN-INDUCED REACTIONS

A few classes of compounds can be oxidized not only by free radical chain reactions, but also by a nonradical process initiated by molecular oxygen. Ordinary, ground-state oxygen, as mentioned above, is a diradical whose structure is approximately $\cdot O-O\cdot$. As such, it reacts readily with free radicals, especially those whose spin density is localized on carbon atoms. However, it is not reactive enough with spin-paired molecules to act as an initiator in the ordinary sense; that is, the reaction

$$RH + \cdot O-O\cdot \rightarrow R\cdot + HOO\cdot$$

(Equation 2-33)

is far too slow to be significant at ordinary temperatures. Radical species like $HO\cdot$, $Cl\cdot$, and even $ROO\cdot$ are much better initiators.

There is another form, singlet oxygen, $^1O_2$, that has a different electronic configuration (roughly $O=O$), without unpaired electrons; that is, it is not a free radical. (Actually, there are two quantum mechanically distinguishable forms of $^1O_2$, but only the lower or $^1\Delta_g$ state is important in solution.) Singlet oxygen is produced by energy transfer from photochemically excited donor molecules, called sensitizers, to ground-state molecular oxygen, which is a triplet (or $^3O_2$) in quantum chemical language. Active sensitizers include some synthetic dyes, such as rose bengal or methylene blue; some environmental pollutants, such as polycyclic hydrocarbons; and some naturally occurring substances, such as chlorophyll and humic materials (Schaap, 1976).

The product of the photoreaction, $^1O_2$, is more reactive than ground-state O_2 toward several environmentally and biologically important compounds. One of its most important interactions with other molecules, however, is collisional quenching (see also Chapter 3). In this process $^1O_2$ is converted back to $^3O_2$ and the acceptor molecule dissipates the collisional energy by passing to a thermally excited state. This type of quenching occurs with (among other compounds) aliphatic amines, which efficiently deactivate $^1O_2$ without undergoing chemical change (Bellus, 1979; Encinas et al., 1987).

Only a few classes of compounds react readily to form new chemical substances, usually by ene and diene reactions, with $^1O_2$:

(a) Olefins, with the most electron-rich varieties reacting fastest. The rate of reaction is also increased when multiple or conjugated sites of unsaturation are present. The naturally occurring compound that is most reactive toward $^1O_2$ is generally agreed to be β-carotene (2-2). Its second-order rate constant for the reaction is about 1×10^{10} l/mol sec, close to the diffusion-controlled limit (Foote and Denny, 1968). Little of the quenching is due to chemical reaction; almost all of it is generally accepted to be physical, or collisional.

(2-2)

The principal products of the chemical reactions of olefins are hydroperoxides, the result of an ene reaction in which the oxygen molecule adds to one end of the original double bond, which then shifts into an allylic position:

$$>C=C-CR_2- + {}^1O_2 -> >\overset{\overset{OOH}{|}}{C}-C=CR_2-$$

(b) Polycyclic aromatic hydrocarbons, exemplified by rubrene (2-3), an orange hydrocarbon that forms a colorless, cyclic endoperoxide by addition of $^1O_2$ in Diels-Alder (4 + 2 cycloaddition) fashion (Monroe, 1978):

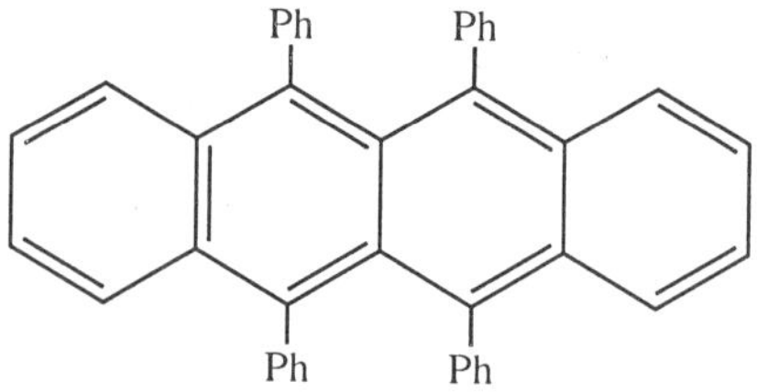

[Structure 2-3: rubrene - tetracene with four Ph substituents]

(2-3)

(c) Cyclic dienes, including heterocyclic dienes such as furans and pyrroles as well as carbocyclic dienes. These compounds also usually react initially by the Diels-Alder route, producing endoperoxides that are usually unstable to further thermal reactions. For example, furfuryl alcohol (2-4) reacts with $^1O_2$ with a rate constant of 1.2×10^8 l/mol sec to form several oxygenated products (Equation 2-34; Haag et al., 1984):

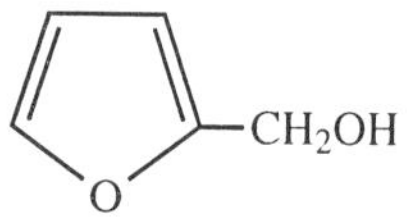

[Structure 2-4: furan with CH₂OH substituent]

(2-4)

[Equation 2-34: furfuryl alcohol + $^1O_2$ → three oxygenated products]

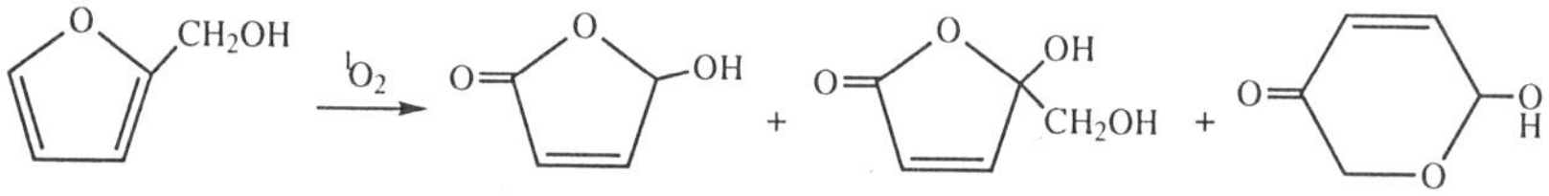

(Equation 2-34)

(d) Sulfur compounds undergo rapid reactions with sensitizing dyes, but they have not been fully characterized mechanistically. Reactions with the triplet state of the sensitizer as well as $^1O_2$ reactions are likely. The products include sulfoxides and disulfides.

(e) Phenols, especially those with multiple alkyl substitution or those that are otherwise electron-rich (e.g., catecholamines, polyphenols), form cyclohexadienone hydroperoxides with $^1O_2$. Again, the mechanism appears to be a preliminary Diels-Alder addition followed by rearrangement (Thomas and Foote, 1978). Phenolate anions are much more electron-rich than the parent phenols due to the increased electron-donating ability of the phenoxide substituent, and at relatively high pHs the contribution of their faster reactions with $^1O_2$ may become important (Scully and Hoigné, 1987). Other extremely electron-rich aromatic compounds, such as polyalkoxylated benzenes (Saito et al., 1972) and polyalkylated naphthalenes (Sakugari et al., 1982) are also reactive toward $^1O_2$.

Singlet oxygen has been repeatedly mentioned as a cause or intermediate of photoinduced toxicity to living organisms. The *in vivo* evidence for it, however, is not usually very strong (Held and Hurst, 1978; Packer et al., 1981; Kanofsky, 1989). Certainly on chemical grounds singlet oxygen is a selective species, and its rates of reaction with most biological substrates are very low. Furthermore, it is quenched back to the ground state by collisions with solvent molecules. The quenching mechanism is particularly important in environments high in water, since the lifetime of $^1O_2$ in that solvent is extremely short (only 4 μsec: Rodgers and Snowden, 1982). Alternative mechanisms of phototoxicity should be considered by photobiologists; in particular, electron transfer from photoexcited molecules to biological targets seems to occur in many systems (Larson and Marley, 1994).

Among the potential reactive sites for $^1O_2$ attack in cells, a few protein amino acids (cysteine, methionine, tryptophan, tyrosine, and histidine) react with it at reasonably rapid rates. Their rate constants, however, are in the 10^7 l/mol sec range, meaning that they are 2–3 orders of magnitude less reactive than β-carotene. (The other amino acids react, if at all, with rate constants at least another three orders of magnitude lower). Histidine and tyrosine show pH dependence on reactivity; histidine is photooxidized only at pHs where the imidazole group is not ionized, whereas tyrosine is oxidized only at pHs where the phenolic group is ionized (Spikes and Straight, 1967).

The products of amino acid oxidation with $^1O_2$ are not very well understood. Methionine sulfoxide is the first product of sensitized methionine photolysis, but other compounds accumulate after longer irradiation times. Histidine, tryptophan, and tyrosine afford complex mixtures of products associated with cleavage of their ring systems (Michaeli and Feitelson, 1994). Tryptophan, one of the most reactive amino acids toward $^1O_2$ (rate constant about 5×10^7 l/mol sec), may possibly disappear from marine waters by a sensitized pathway (Momzikoff et al., 1983), although tryptophan does also absorb some solar UV and could be destroyed directly.

Unsaturated fatty acids also react, but at much slower rates. Molecules of this class that have more carbon–carbon double bonds are more reactive: for example, linolenic acid (3 double bonds) has a rate constant of 3×10^5 l/mol sec; linoleic acid (2 double bonds), 2×10^5 l/mol sec; and oleic acid (one double bond), 1×10^5 l/mol sec (Wilkinson and Brummer, 1981). The reaction with these compounds, like that with other olefins, is formally an ene reaction. The initial product of the reaction is a lipid hydroperoxide;

$$-CH=CH-CH_2-CH=CH- \rightarrow -CH-CH=CH-CH=CH- \\ | \\ OOH$$

(Equation 2-35)

Breakdown products of hydroperoxides include alcohols, ketones, epoxides, and cross-linked oligomers and polymers (Girotti, 1990). Unsaturated lipids also are attacked by free radical species to give similar product mixtures, but in these cases the peroxyl radical initiates the process.

Among nucleic acid derivatives, only guanine-containing compounds show significant reactions. Guanine itself (rate constant about 5×10^6 l/mol sec) undergoes specific hydroxylation at the 8-position to give 2-5:

(2-5)

This product was once suggested as a marker for $^1O_2$-initiated damage, but subsequent work has clearly shown that it is formed by many other oxidants (Devasagayam et al., 1991; Simic, 1994; Rosen et al., 1996).

Singlet oxygen-induced damage in tissues could theoretically be minimized either by preventing the $^1O_2$ from forming (intercepting the excited states of sensitizers) or reacting rapidly with $^1O_2$ once it is formed (Davidson and Trethewy, 1976). Because the sensitizer concentration would be expected to be much higher than the $^1O_2$ steady-state concentration (at least in an environment high in water), the former mechanism ought to be quite important. Within cells, both mechanisms are probably co-occurring. Many quenchers of excited states, such as β-carotene, are also potent quenchers of $^1O_2$.

III. OTHER MECHANISMS OF OXIDATIVE DAMAGE IN CELLS AND TISSUES

Despite the considerable attention paid to singlet oxygen mechanisms of biological damage, it is probable that radical pathways are more important. Reactive, oxidizing species derived from oxygen are formed in cells by normal metabolism as well as by environmental insults. In this section the major types of biologically important free radical mechanisms will be briefly surveyed.

A. Superoxide

Wherever molecular oxygen is present in a system where free electrons are being formed, superoxide is generated. This may include aquatic or aerobic environments where sunlight impinges on natural organic matter, or living

AUTOOXIDATION MECHANISMS

tissue where redox-active enzymes generate reducing conditions. Superoxide is produced by many enzymes (oxidases, dehydrogenases, and hydroxylases) characteristic of diverse tissues. Xanthine oxidase was the first of these enzymes to be described and probably still the most effective producer of $\cdot O_2^-$ among known enzymes (McCord and Fridovicl., 1968). Whole cells, particularly those involved in defense against infection (such as leukocytes) also release superoxide into the medium that surrounds them (Fridovich, 1974). Superoxide is also produced during the direct one-electron air oxidation of transition metal ions such as iron(II). In these oxidations, atmospheric molecular oxygen forms a complex with the ion, and electron transfer occurs within the complex (Uri, 1961). This pathway is probably important in natural aquatic environments, as well as in cells where iron complexes such as hemoglobin take part in oxygen metabolism.

The superoxide radical anion, $\cdot O_2^-$ exists in equilibrium with its conjugate acid, the perhydroxyl radical HOO·, which has a pKa of about 4.8 (Bielski and Allen, 1977). The perhydroxyl radical plays important roles in atmospheric chemistry, where because of the slightly acidic character of most cloudwaters it is the dominant oxygen free radical present. A number of reports have also indicated that HOO· is more reactive with biologically important molecules than $\cdot O_2^-$; for example, HOO· fosters the oxidation of polyunsaturated fatty acids by removing either their bis-allylic hydrogen atoms or the protons of lipid hydroperoxides (Bielski et al., 1983; Aikens and Dix, 1991):

$$[-CH=CH-CH_2-CH=CH-] + HOO\cdot \rightarrow [-CH=CH-CH\cdot-CH=CH-] \rightarrow \rightarrow$$

$$[-CH=CH-\overset{\overset{OOH}{|}}{CH}-CH=CH-] + HOO\cdot \rightarrow$$

$$[-CH=CH-\overset{\overset{OO\cdot}{|}}{CH}-CH=CH-]$$

(Equation 2-36)

Various other important compounds (ascorbate, phenols, amino acids) have also been shown to be more readily attacked by HOO· than $\cdot O_2^-$ (Bielski and Shue, 1979; Bielski, 1983).

The anion of superoxide, by contrast, is not of itself a potent oxidant, although it does oxidize thiols at pHs where the HOO· species should be present at insignificant concentrations (Asada and Kanematsu, 1976):

$$RSH + \cdot O_2^- \rightarrow RS\cdot$$

(Equation 2-37)

This reaction, and those of other "superoxide-oxidizable" compounds such as polyphenols, ascorbate, etc., may, however, not be simple electron-transfer oxidations. An alternative mechanism has been suggested by Nanni et al. (1980) and refined by Sawyer et al. (1985). Superoxide rapidly removes both protons and hydrogen atoms from these substrates; the corresponding anions and radicals are then much more susceptible to oxidation by the products of superoxide dismutation, H_2O_2 and O_2. The reaction is illustrated here for catechol:

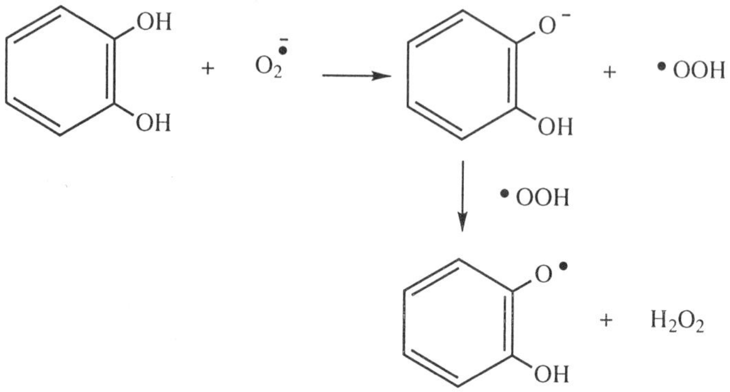

$$\text{catechol(OH,OH)} + O_2^{\bullet -} \longrightarrow \text{catechol}(O^-, OH) + {}^{\bullet}OOH$$

$$\downarrow {}^{\bullet}OOH$$

$$\text{catechol}(O^{\bullet}, OH) + H_2O_2$$

(Equation 2-38)

Superoxide also reacts rapidly with nitric oxide, NO, a metabolic intermediate playing important roles in cellular regulation and communication processes, to form another potent oxidizing agent, peroxynitrite ($ONOO^-$) (Koppenol et al., 1992). This substance has been demonstrated to oxidize lipids and proteins, to cause DNA strand breaks, and to form oxidized derivatives of the DNA base, guanine (Pryor and Squadrito, 1995; Douki and Cadet, 1996).

B. Hydrogen Peroxide

The sources of hydrogen peroxide in the environment, and in living cells, potentially include all reactions where superoxide is produced, since superoxide disproportionates so rapidly to H_2O_2 and oxygen (except in highly alkaline environments). This includes the enzymatic processes outlined above, as well as photoionization, and also the one-electron autooxidation of transition metals, such as the air oxidation of iron(II) to iron(III).

Superoxide is decomposed by superoxide dismutase (SOD) in aerobic organisms (Fridovich, 1974, 1995). The products of the reaction are oxygen and hydrogen peroxide (Equation 2-39). Hydrogen peroxide may have some normal metabolic functions, but more typically it is thought of as a potentially damaging oxidizing agent that requires control,

$$2 \cdot O_2^- + 2H^+ \xrightarrow{SOD} H_2O_2 + O_2$$

(Equation 2-39)

a cellular function that is usually provided by the enzyme catalase, which breaks H_2O_2 down into water and molecular oxygen. In fact, however, H_2O_2 by itself is a rather weak oxidant and most organic compounds (except for some sulfur-containing molecules) are virtually inert to attack by it at ordinary environmental or cellular concentrations and temperatures. In the presence of reduced transition metal ions, however, H_2O_2 is converted to the much more reactive oxidant, hydroxyl radical:

$$H_2O_2 + Fe^{2+} \rightarrow Fe^{3+} + HO^- + HO\cdot$$

(Equation 2-40)

The metal ion in Equation 2-40, often referred to as the Fenton reaction, need not be iron; studies have shown that copper(I), cobalt(II) and nickel(II) also take part in the process (Halliwell and Gutteridge, 1990).

Other reactive oxidants, including various ill-defined complexes containing higher valence states of iron, may also be transiently formed in the Fenton reaction under certain conditions. In fact, some investigators of the Fenton reaction have suggested that HO· is not even an intermediate (Sawyer et al., 1993). It is therefore not surprising to learn that organisms maintain low intracellular free iron concentrations by securing it in stable complexes like porphyrins and in iron storage proteins such as ferritin (Weinberg, 1990). However, in conditions where mammalian cells become damaged, their contents (including iron-containing proteins) are discharged into the surrounding tissues, where thay may induce further damage by taking part in free-radical reactions (Halliwell and Gutteridge, 1985).

Hydrogen peroxide is deliberately used as an oxidizing reagent by most organisms, in conjunction with enzymes known as peroxidases (Gardner, 1980; Everse et al., 1991; Reddy and D'Souza, 1994). These enzymes contain transition-metal (usually iron) cofactors, and therefore presumably generate much more reactive oxidants. Sometimes activated in damaged or wounded cells, one of their important functions seems to be the removal of defective or damaged cellular materials. Other organisms, such as wood-degrading fungi, use them as extracellular oxidizing agents for recalcitrant substances such as lignin.

C. Hydroxyl Radical

The hydroxyl radical is such a potent oxidizing agent that it has very little discrimination among potential substrates for attack. With electron-rich compounds like phenols and anilines, HO· reacts at virtually diffusion-controlled rates (ca. $10^9 - 10^{10}$ l/mol sec), but even with organic compounds not usually thought of as susceptible to oxidation, such as n-alkanes, it reacts rapidly at rates only about two orders of magnitude slower. Only a few biologically important molecules are (more or less) unreactive with HO·, notably urea, whose second-order rate constant is "only" about 10^5 l/mol sec. Accordingly, it is expected that HO·, if formed within a cell or at a cell surface, would react within a very few molecular diameters of the site of its generation, that is, it would have a small "reaction volume" (a property of reactive transients that depends on their diffusivity and lifetime). Therefore, it is difficult to conceive of biologically reasonable cases where an antioxidant could exist in high enough concentrations to scavenge HO· in a cell.

Hydroxyl radicals have two principal modes of reaction with organic compounds, hydrogen atom (or electron) abstraction, and (somewhat faster) addition to double bonds. The two processes may be illustrated by considering the products of reaction of HO· with toluene, which include o-, m-, and p-cresols, dimethylbiphenyls, and bibenzyl, as well as products of further oxidation (Hoshino et al., 1978). In each case, a secondary free radical species is produced, in keeping with the truism "odd + even = odd," and it is this secondary radical that is likely to be important in most instances of HO·-induced damage in cells.

When the hydroxyl radical attacks a C–H bond, the reaction product is a carbon-centered free radical, $R_3C·$, typically reactive with molecular oxygen at extremely fast, almost diffusion-controlled, rates (see Section A-2 of this chapter). Because of the diradical character of oxygen molecules, there is little or no activation energy (spin barrier) for such an odd-odd electron pairing reaction. The reaction products, peroxyl radicals, ROO·, are probably extremely important intermediates in many biological processes such as lipid peroxidation. The general characteristics of reactions of ROO· have been described above (Section A-3).

REFERENCES

Aikens, J. and T. A. Dix. 1991. Perhydroxyl radical (HOO·) initiated lipid peroxidation. J. Biol. Chem. 266, 15091–15098.

Asada, K. and S. Kanemitsu. 1976. Reactivity of thiols with superoxide radicals. Agr. Biol. Chem. 40, 1891–1892.

Bellus, D. 1979. Physical quenchers of molecular oxygen. Adv. Photochem. 11, 105–205.

Bielski, B. H. J. 1983. Evaluation of the reactivities of HO_2/O_2^- with compounds of biological interest. In G. Cohen and R. A. Greenwald, eds., *Oxyradicals and Their Scavenger Systems.* Vol. 1, Molecular aspects. Elsevier, New York.

Bielski, B. H. J., and A. O. Allen. 1977. Mechanism of the disproportionation of superoxide radicals. J. Phys. Chem. 81, 1048–1050.

Bielski, B. H. J. and G. G. Shiue. 1979. Reaction rates of superoxide radicals with the essential amino acids. In *Oxygen Free Radicals and Tissue Damage* (CIBA Foundation symposium #65), Excerpta Medica, Amsterdam. pp. 43–56.

Bielski, B. H. J., R. L. Arudi, and M. W. Sutherland. 1983. A study of the reactivity of HO_2/O_2^- with unsaturated fatty acids. J. Biol. Chem. 258, 4759–4761.

Burton, A., K. U. Ingold, and J. C. Walton. 1996. Absolute rate constants for the reactions of primary alkyl radicals with aromatic amines. J. Org. Chem. 61, 3778–3782.

Devasagayam, T. P. A., S. Steenken, M. S. Oberdorf, W. A. Schultz, and H. Sies. 1991. Formation of 8-hydroxy(deoxy)guanosine and generation of strand breaks at guanine residues in DNA by singlet oxygen, Biochemistry 30, 6283–6289.

Douki, T. and J. Cadet. 1996. Peroxynitrite mediated oxidation of purine bases of nucleosides and isolated DNA. Free Rad. Res. 24, 369–380.

Encinas, M. V., E. Lemp, and E. A. Lissi. 1987. Interaction of singlet oxygen with aliphatic amines and hydroxylamines. J. Chem. Soc. Perkin Trans. II, 1125–1127.

Eriksen, J., C. S. Foote, and T. L. Parjer. 1977. Photosensitized oxygenation of alkenes and sulfides by a non-singlet oxygen mechanism. J. Am. Chem. Soc. 99, 6455–6456.

Everse, J., K. E. Everse, and M. B. Grisham, eds. 1991. *Peroxidases in Chemistry and Biology.* CRC Press, Boca Raton, FL.

Foote, C. S. and R. W. Denny. 1968. Chemistry of singlet oxygen. VII. Quenching by β-carotene, J. Am. Chem. Soc. 90, 6233–6235.

Franz, J. A., D. H. Roberts, and K. F. Ferris. 1987. Absolute rate expressions for intramolecular displacement reactions of primary alkyl radicals at sulfur. J. Org. Chem. 52, 2256–2262.

Fridovich, I. 1974. Superoxide dismutases. Adv. Enzymol. 41, 35–97.

Fridovich, I. 1995. Superoxide radical and superoxide dismutases. Ann. Rev. Biochem. 64, 97–112.

Furimsky, E., J. A. Howard, and J. Selwyn. 1980. Absolute rate constants for hydrocarbon autoxidation. 28. A low temperature kinetic spin resonance study of the self-reactions of isopropylperoxyl and related secondary alkylperoxyl radicals in solution. Can. J. Chem. 58, 677–680.

Gardner, H. W. 1980. Lipid enzymes: lipases, lipoxygenases, and "hydroperoxidases." In M. G. Simic and M. Karel, eds., *Autoxidation in Food and Biological Systems,* Plenum, New York, pp. 447–504.

Girotti, A. W. 1990. Photodynamic lipid peroxidation in biological systems. Photochem. Photobiol. 51, 497–509.

Haag, W. R., J. Hoigné, E. Gassmann, and A. M. Braun. 1984. Singlet oxygen in surface waters. 1. Furfuryl alcohol as a trapping agent. Chemosphere 13, 631–640.

Halliwell, B. and J. M. C. Gutteridge. 1985. Oxygen radicals and the nervous system. Trends Neurosci. 8, 22–26.

Halliwell, B. and J. M. C. Gutteridge. 1990. Role of free radicals and catalytic metal ions in human disease: an overview. Meth. Enzymol. 186, 1–85.

Held, A. M. and J. K. Hurst. 1978. Ambiguity associated with use of singlet oxygen trapping agents in myeloperoxidase-catalyzed oxidations. Biochem. Biophys. Res. Commun. 81, 878–885.

Hendry, D. G., T. Mill, L. Piskiewicz, J. A. Howard, and H. K. Eigenmann. 1974. A critical review of H-atom transfer in the liquid phase: chlorine atom, alkyl, trichloromethyl, alkoxy, and alkylperoxy radicals. J. Phys. Chem. Ref. Data 3, 937–978.

Hoshino, M., H. Akimoto, and M. Okuda. 1978. Photochemical oxidation of benzene, toluene, and ethylbenzene initiated by OH radicals in the gas phase. Bull. Chem. Soc. Japan 51, 718–724.

Kanofsky, J. R. 1989. Singlet oxygen production by biological systems. Chem.-Biol. Interact. 70, 1–28.

Koppenol, W. H., J. J. Moreno, W. A. Pryor, H. Ischiropolus, and J. S. Berckman. 1992. Peroxynitrite, a cloaked oxidant formed by nitric oxide and superoxide. Chem. Res. Toxicol. 5, 834–842.

Larson, R. A. and K. A. Marley. 1994. Oxidative mechanisms of phototoxicity. In J. O. Nriagu and M. S. Simmons, eds. *Environmental Oxidants* (Advances in Environmental Science and Technology). John Wiley and Sons, New York. pp. 269–317.

Mattay, J. 1987. Charge transfer and radical ions in photochemistry. Angew. Chem. Int. Ed. 26, 825–845.

Mattes, S. L. and S. Farid. 1982. Photochemical cycloadditions via exciplexes. Acc. Chem. Res. 15, 80–86.

McCord, J. M. and I. Fridovich. 1968. The reduction of cytochrome c by milk xanthine oxidase. J. Biol. Chem. 243, 5753–5760.

Michaeli, A. and J. Feitelson. 1994. Reactivity of singlet oxygen toward amino acids and peptides. Photochem. Photobiol. 59, 284–289.

Monroe, B. M. 1978. Rates of reaction of singlet oxygen with olefins. J. Phys. Chem. 82, 15–18.

Nanni, E. J., M. D. Stallings, and D. T. Sawyer. 1980. Does superoxide ion oxidize catechol, α-tocopherol, and ascorbic acid by direct electron transfer? J. Am. Chem. Soc. 102, 4481–4485.

Neta, P., R. E. Huie, and A. B. Ross. 1990. Rate constants for reactions of peroxyl radicals in fluid solutions. J. Phys. Chem. Ref. Data 19, 413–513.

Neta, P., J. Grodkowski, and A. B. Ross. 1996. Rate constants for the reactions of aliphatic carbon-centered radicals in aqueous solution. J. Phys. Chem. Ref. Data 25, 709–1050.

Nieman, M. 1964. Mechanism of the oxidative thermal degradation and of the stabilization of polymers. Russ. Chem. Rev. 33, 13–27.

Packer, J. E., J. S. Mahood, V. O. Mora-Arellano, T. F. Slater, R. L. Willson, and B. S. Wolfenden. 1981. Free radicals and singlet oxygen scavengers: reaction of a peroxy-radical with β-carotene, diphenylfuran, and 1,4-diazabicylco[2.2.2]octane. Biochem. Biophys. Res. Commun. 98, 901–906.

Pryor, W. A. and G. L. Squadrito. 1995. The chemistry of peroxynitrite, a product from the reaction of nitric oxide with superoxide. Am. J. Physiol. 12, L699–L722.

Reddy, C. A. and T. M. D'Souza. 1994. Physiology and molecular biology of the lignin peroxidases of *Phanaerochaete chrysosporium*. FEMS Microbiol. Revs. 13, 137–152.

Rosen, J. E., A. K. Prahalad, and G. M. Williams. 1996. 8-Oxoguanosine formation in the DNA of cultured cells after exposure to H_2O_2 alone or with UVB or UVA irradiation. Photochem. Photobiol. 64, 117–122.

Russell, G. A. 1957. Deuterium-isotope effects in the autoxidation or aralkyl hydrocarbons. Mechanism of the interaction of peroxy radicals. J. Am. Chem. Soc. 79, 3871–3877.

Rånby, B. and J. F. Rabek. 1975. *Photodegradation, Photo-Oxidation and Photostabilization of Polymers*. Wiley, London.

Reich, L. and S. Stivala. 1969. *Autoxidation of Hydrocarbons and Polyolefins: Kinetics and Mechanisms.* Marcel Dekker, New York.

Sawyer, D. T., T. S. Calderwood, C. L. Johlman, and C. L. Wilkins. 1985. Oxidation by superoxide ion of catechols, ascorbic acid, dehydrophenazine, and reduced flavins to their respective anion radicals. A common mechanism via a sequential proton-hydrogen atom transfer. J. Org. Chem. 50, 1409–1412.

Sawyer, D. T., C. Kang, A. Llobet, and C. Redman. 1993. Fenton reagents (1:1 $Fe^{III}L_x/HOOH$) react via $[L_x Fe^{II}OOH(BH^+)]$ (1) as hydroxylases (RH → ROH), not as generators of free hydroxyl radicals (HO·). J. Am. Chem. Soc. 115, 5817–5818.

Schaap, A. P., ed. 1976. *Singlet Molecular Oxygen.* Dowden, Hutchinson, and Ross, Stroudsburg, PA.

Simic, M. G. 1994. DNA markers of oxidative processes *in vivo*: relevance to carcinogenicity and anticarcinogenicity. Cancer Res. 54, S1918–S1923.

Simic, M. G. and E. P. L. Hunter. 1983. Interaction of free radicals and antioxidants. In O. F. Nygaard and M. G. Simic, eds., *Radioprotectors and Anticarcinogens.* Academic Press, New York, pp. 449–460.

Spikes, J. D. and R. Straight. 1967. Sensitized photochemical processes in biological systems. Ann. Rev. Phys. Chem. 18, 409–436.

Uri, N. 1961. Physico-chemical aspects of autoxidation. In W. O. Lindberg, ed., *Autoxidation and Antioxidants.* Wiley-Interscience, New York. Vol. 1, pp. 55–106.

von Sonntag, C. 1987. *The Chemical Basis of Radiation Biology.* Taylor and Francis, London.

Weinberg, E. D. 1990. Cellular iron metabolism in health and disease. Drug Metab. Rev. 22, 531–579.

Wilkinson, F. and J. G. Brummer. 1981. Rate constants for the decay and reactions of the lowest electronically excited singlet state of molecular oxygen in solution. J. Phys. Chem. Ref. Data 10, 809–999.

Willson, R. L. 1985. Organic peroxy free radicals as ultimate agents in oxygen toxicity. In H. Sies, ed., *Oxidative Stress.* Academic Press, London, pp. 41–72.

3 QUENCHING AND SCAVENGING OF REACTIVE SPECIES

I. RADICAL SCAVENGING AND REDOX POTENTIAL

Almost any organic substance is subject to gradual, or not-so-gradual, decomposition in the presence of oxygen. Living things have evolved many defense mechanisms to prevent their tissues from autooxidation; mankind, too, has devised numerous clever techniques to prevent manufactured objects from becoming useless in a brief period. Since autooxidative decomposition is principally a free-radical process, as described in Chapters 1 and 2, protective agents that destroy or inactivate radicals make up a primary line of defense.

Free radicals that may cause damage or dysfunction in either nonliving or living systems could presumably be prevented from exerting their harmful effects by several means. Either physical and chemical techniques could, in principle, be employed to limit the potential damage.

A. Preventive Antioxidation

First, measures could be taken to prevent the free radicals from being formed at all. A substance capable of being attacked by oxygen could be coated with a resistant material; or photoinduced autooxidation could be inhibited by adding a shielding or reflective layer; or the material could be stored or used in an oxygen-free environment. This approach has been termed "preventive (or "obstructive") antioxidation." An example of this strategy might be the covering of a wood surface with an opaque layer of paint, that would prevent light and damaging oxygen-derived species from gaining access to the structural biopolymers of the wood. Analogously, rubber objects such as tires are treated with carbon black or other light-scattering materials to prevent the penetration of damaging wavelengths into the object.

Living organisms employ similar strategies. One approach is simply to avoid oxygen altogether. Many microorganisms live in environments that are either totally anoxic or limited in oxygen concentration. Other life forms eschew sunlight and occupy permanently dark environments, such as the ocean depths, subsurface layers of the soil, or caves. The surfaces of many animals that do live in the presence of sunlight and oxygen are either dark in color, highly reflective, or both, presumably at least in part because of the potentially toxic effects of light. An interesting example of the latter phenomenon is a melanic (brown to black) mutant of the tobacco hornworm (*Manduca sexta*). This animal is a moth larva that feeds on the foliage of plants such as the tomato that are known to contain organic compounds that become toxic when exposed to sunlight. The dark-colored form is much more resistant to the toxic effects of these compounds (Berenbaum, 1987).

Some authors also use the term "preventive autooxidation" to refer to the mechanism of action of chemical additives that inhibit the formation of chain initiating species such as singlet oxygen or hydroperoxides. They prefer to reserve the term "chain-breaking antioxidants" for compounds that specifically react with the free-radical intermediates. such as R· and ROO·, of the typical chain reaction. (In quite a few cases, it is not always clear by which mechanism an antioxidant may be functioning.) The reader of the literature of the field should keep these different perspectives in mind.

B. Chemical Antioxidation

The most common and useful approach used in preventing damage to autooxidizable materials is to incorporate chemical additives in the formulation to deactivate the species that initiate or promote destructive oxidation reactions. As will be recalled from earlier chapters, autooxidation reactions are normally initiated by species capable of producing free radicals, which then undergo rapid subsequent reactions with molecular oxygen leading to damage. Protective additives might be light-absorbing compounds (materials that prevent significant levels of UV radiation from entering the substance being protected); metal ion-complexing agents; free radical scavengers; peroxide-destroying compounds; or singlet oxygen quenchers.

The majority of autooxidation initiators and promoters are oxidizing agents. As such, they should react most readily with strong electron donors (easily reducible compounds). In principle, one should be able to predict which chemical species would be the best antioxidants by examining some measurable physical property such as redox potential.

Thermodynamically, the ability of radical scavengers to intercept oxidizing free radicals should depend at least to some extent on their relative one-electron redox potentials. This parameter is a measure of the ability of chemical species to be oxidized or reduced by external odd-electron species. Obviously, the thermodynamic potentials will never tell the whole story, since the overcoming of transition-state kinetic barriers to reaction invariably complicate such oversimplifying considerations. Furthermore, steric, solvent and solubility, and interfacial effects all enter into the overall picture. Nevertheless, redox considerations are at least a starting point for antioxidant analysis.

Free radicals exhibit a tremendous range of redox potentials; some are highly oxidizing, and some highly reducing. Table 3-1 lists some selected data on the one-electron redox potentials (in aqueous solution at pH 7.0) of some species that are of biochemical or environmental interest. Various methods have been used to calculate or measure these potentials (for example, equilibria or other thermodynamic measurements, direct electrochemical cell techniques, flash photolysis and pulse radiolysis, and comparison to agents with known potentials), and values may vary greatly from one compilation to another. Still, this table is a starting point for general considerations of antioxidant potency, at least toward free-radical oxidants. (Some of the redox chemistry of major classes of naturally occurring antioxidants will be discussed in increased detail in later chapters.)

The data in the table may be used to predict the likelihood of a redox reaction taking place between an oxidized form (listed in the left-hand column) and a reduced form from the second column. Other things being equal, any oxidized species should be capable of accepting an electron from a reduced species lower in the table. The implication would be that the most effective antioxidants should be those substances that are farthest down the list in the right-hand column. This is basically true as far as it goes, but it should be remembered that living cells are never uncomplicated aqueous solutions with a few simple species present at well-defined concentrations, and neither, for that matter, are most environmental milieux. Furthermore, for a redox reaction actually to take place at a measurable rate, there must be a significant concentration of the active species present at the site of the reaction. This clearly is a condition that cannot be satisfied for alkali metals in aqueous solution, or reactive free radicals almost anywhere. Therefore, it is necessary for there to be a mechanism for keeping the potential antioxidant in its reduced form, delivering it to the site where oxidative damage is occurring and, if it is consumed during the reaction, to restore or replenish its activity. These are rather stringent requirements that greatly limit the number of useful antioxidants.

II. METAL COMPLEXATION AND INHIBITION OF RADICAL REACTIONS

Many damaging autooxidation reactions are initiated by metal ions, especially oxidizing transition metal ions having stable valence forms that differ by one electron from a reduced form. Examples include Fe(III), Cu(II), and Co(III). Even very low levels ($<10^{-8}$ M) of active metal species can have significant effects on the degradation of readily oxidizable compounds. It should be kept in mind that these levels of metal contamination are not uncommon in laboratory reagents such as buffer salts (Buettner, 1986).

Table 3-1 Redox Potentials (1-Electron) of Environmentally and Biochemically Significant Compounds and Elements at pH 7.0 and 25°C (unless indicated)

Oxidized form	Reduced form	Potential (mv)	Ref #	Notes
$F\cdot$	F^-	3600	10	
$\cdot SO_4^-$	SO_4^{2-}	2430	10	
$Cl\cdot$	Cl^-	2410	10	
$HO\cdot$	H_2O	2310	1	
$CH_3CH_2\cdot$	Ethane	1900	1	
Co(III)	Co(II)	1820	5	
$\cdot O_3^-$	$H_2O + O_2$	1800	1	
Ce(IV)	Ce(III)	1610	5	
$RO\cdot$	ROH	1600	1	Aliphatic
HOCl	½ Cl_2	1590	4	
$\cdot CO_3^-$	CO_3^{2-}	1500	10	
$N_3\cdot$	N_3^-	1330	1	
$\cdot CH_2OH$	Methanol	1200	1	
Fe(III) phenanthr	Fe(II) phenanthr	1150	1	
$HOO\cdot$	H_2O_2	1060	1	
$ROO\cdot$	ROOH	1000	1	Alkyl R: ranges from 770–1400
$\cdot O_2^-$	H_2O_2	940	1	
Tyrosine-$O\cdot$	Tyrosine	940	8	
ClO_2	ClO_2^-	930	10	
$RS\cdot$	RSH	920	1	Cysteine
$C(NO_2)_4$	$\cdot C(NO_2)_3, NO_2^-$	900	10	
$PhO\cdot$	PhOH	900	1	At pH 5.6, ranges from 780–1170
O_3	O_3^-	890	1	
O_2 (singlet)	$\cdot O_2^-$	830	10	
Cl_2	Cl_2^-	700	10	
Serotonin-$O\cdot$	Serotonin	640	7	
Sesamol-$O\cdot$	Sesamol	620	9	
$PUFA\cdot$	PUFA	600	1	Bis-allylic H
$Urate^-\cdot$	Uric acid	590	1	
Br_2	Br_2^-	580	10	
Catechol-$O\cdot$	Catechol	530	1	
Tocopherol-$O\cdot$	Tocopherol	500	1	Vitamin E
Trolox-$O\cdot$	Trolox	480	1	
Benzosemiquinone	Hydroquinone	460	3	
H_2O_2	H_2O	320	1	
$Ascorbate^-\cdot$	$Ascorbate^-$	282	1	Vitamin C
Ferricytochrome C	Ferrocytochrome C	260	1	
$\cdot SO_3$	SO_3^-	250	10	
I_2	I_2^-	210	10	
Ubisemiquinone	Ubiquinol	200	1	Coenzyme Q
Cu(II)	Cu(I)	160	4	
Fe(III) EDTA	Fe(II) EDTA	120	1	
Fe(III) aq	Fe(II) aq	110	1	770 at pH 1
Benzoquinone	Benzosemiquinone	100	3	

Table 3-1 Redox Potentials (1-Electron) of Environmentally and Biochemically Significant Compounds and Elements at pH 7.0 and 25°C (unless indicated) *(Continued)*

Oxidized form	Reduced form	Potential (mv)	Ref #	Notes
Fe(III) DETAPAC	Fe(II) DETAPAC	30	1	
Hg(II)	Hg(I)	0	10	
Ubiquinone	Ubisemiquinone	−36	1	
Dehydroascorbate	Ascorbate−·	−174	1	
Fe(III) ferritin	Fe(II) ferritin	−190	1	
Duroquinone	Durosemiquinone	−264	1	
Riboflavin	Riboflavin semiquinone	−317	1	
O_2	$·O_2^-$	−330 (−160)	1	(3)
$CFCl_3$	$−CFCl_2$, $Cl·$	−440	10	
Paraquat	Paraquat−·	−448	1	
Fe(III) desferal	Fe(II) desferal	−450	1	
O_2	HOO·	−460	1	
$PhNO_2$	$PhNO_2^-·$	−485	3	
CCl_4	$−CCl_3$, $Cl·$	−540	3	
MPP+ (6)	MPP·	−1120	6	
$CHCl_3$	$−CHCl_2$, $Cl·$	−1440	3	
RSSR	RSSR−·	−1500	1	
CO	HCO	−1540	10	
CO_2	$CO_2^-·$	−1800	1	
H+	H·	−2310	10	
Na+	Na	−2710	2	
H_2O	e_{aq}	−2870	1	

From (1) Buettner, G. R., 1993; (2) Stumm, W. and J. J. Morgan, 1981; (3) Schwarzenbach et al., 1993; (4) Pankow, J. F., 1991; (5) Moore, W. J., 1962; (6) Oturan, M. et al., 1988; (7) Jovanovic, S. et al., 1990; (8) Lind, J. et al., 1990; (9) Jovanovic, S. et al., 1991; and (10) Stanbury, D. M., 1989.

Redox reactions in solution in which metallic ions take part are always affected by the ligands that surround the ion. Both promotion and inhibition of, for example, an initiation reaction involving a specific metal ion have been reported. Ligand interactions often markedly affect the redox potential of electron exchange reactions, and steric effects involving bulky ligands may cause the complexed ion to become inaccessible to the redox reaction partner.

Synthetic complexing, coordinating, or chelating agents are widely used in the formulation of antioxidant mixtures for polymers, rubber, and other materials. The most effective complex-forming compounds are those that possess lone pairs of electrons to donate to the metal ion, and that have orbitals arranged in space so that the metal ion's vacant orbital sites are accommodated efficiently. Molecules that have two or more electron-donating sites capable of binding with cations are especially good complexing agents (they are referred to as bidentate, tridentate, etc., depending on the number of such groups). Salts of ethylenediamine tetraacetic acid (EDTA: 3-1) exemplify this type of structure. These complexes feature multicoordinating ligands and nearly strain-free rings that result in a highly stabilized structure.

$$\text{N}\begin{matrix}\nearrow \text{CH}_2\text{COOH}\\ \searrow \text{CH}_2\text{COOH}\end{matrix}$$
$$|$$
$$\text{CH}_2$$
$$|$$
$$\text{CH}_2$$
$$|\ \text{N}\begin{matrix}\nearrow \text{CH}_2\text{COOH}\\ \searrow \text{CH}_2\text{COOH}\end{matrix}$$

(3-1)

Of course, different metal ions differ tremendously in their potential to be complexed. It is not always possible to predict the speciation of metallic ions in the presence of complexing agents having different properties. Generally speaking, however, the ions of transition metals (those having closely spaced d-orbitals) participate most readily in hybridized bonding structures involving nearby ligands (Perrin and Sillen, 1979; Hogfeldt, 1982) and have the highest equilibrium, or stability, constants for complex formation. Among the transition metals, smaller, multiply charged ("hard") cations such as Co(III) and Cr(III) tend to be the strongest centers for coordination.

The ability of a complexing agent to accommodate any particular cation is a function of many factors, including solvent or medium effects, ionic strength, temperature, ligand and cation concentrations, the presence of rival complexing species, and pH. For example, EDTA's four carboxyl groups ionize with pKa's of 2.0, 2.8, 6.2, and 10.3 (Meites, 1963). Therefore, at pHs only a little below neutrality, EDTA rapidly loses its ability to chelate most metal ions. These aspects need to be taken into account when assessing the potential roles of complexing agents in cellular or environmental mixtures.

Naturally occurring metal complexing agents include amino acids, hydroxy acids, polycarboxylic acids, polyphenols, phosphates and phosphatides, and possibly diketones. Many of these classes will be discussed in greater detail in subsequent chapters, but here we will briefly mention of a few types of particular importance.

A. Amino Acids and Peptides

Metal ion binding by amino acids has attracted considerable attention, since metallic cofactors are so important for enzymatic activity (Page and Williams, 1987; Yamauchi, 1995). Complexes are often formed between the free amino group and an S, N, or O donor group on the side chain. The stability constant data for many of these complexes have been critically reviewed (Berthon, 1995; Odoani and Yamauchi, 1996).

Work on peptide binding of amino acids has been stimulated by the discovery of a family of related peptides, "phytochelatins," that are produced by numerous algae, fungi, and higher plants (Rauser, 1995). These compounds are technically metallothienins, that is, small peptides with more or less specific metal-complexing ability. In the case of phytochelatins, almost all discovered so far seem to have the structure 3-2 where n ranges from 1 to 4. These compounds seem to be the principal agents for the complexation of heavy metals at least in higher plants.

$$(\gamma\text{-glutamyl-cysteinyl})_n\text{-glycine}$$

$$(3\text{-}2)$$

Several metal-binding proteins have been identified whose function appears to be to maintain the concentration of reactive transition metals at a very low value in the body The plasma protein transferrin, for example, has a very high affinity for iron (Gutteridge et al., 1982). Numerous other extracellular proteins, such as albumin, ceruloplasmin, and lactoferrin, either complex transition metal irons such as iron or copper, or keep them in an oxidized state that are thermodynamically unreactive with hydroperoxides (Krinsky, 1992).

B. Hydroxy and Polycarboxylic Acids

Because of their importance in metabolism, numerous small naturally occurring carboxylic acids have been examined for their abilities to complex metal ions. Some of the more important of these compounds include lactic (3-3), citric (3-4) and tartaric (3-5) acids. The stability constants for most of these complexes have been recorded (Meites, 1963; Perrin and Sillen, 1979; Alumaa and Pentsub, 1994). Because they often occur at relatively high concentrations in cells and tissues, they may be significant competitors for metals that might otherwise bind to more important biomolecules such as proteins (Graff et al., 1995). Citric acid, for example, occurs in many fruits at 1–8% (50–400 mM) levels; but even in other plant tissues it is often present at more than 1 mM concentrations (Madhavi et al., 1995).

$$(3\text{-}3)$$

$$\begin{array}{c} \text{CH}_2\text{COOH} \\ | \\ \text{HO}-\text{CH}-\text{COOH} \\ | \\ \text{CH}_2\text{COOH} \end{array}$$

(3-4)

$$\begin{array}{c} \text{COOH} \\ | \\ \text{CHOH} \\ | \\ \text{CHOH} \\ | \\ \text{COOH} \end{array}$$

(3-5)

The acids themselves and many of their derivatives have been used as antioxidants in foodstuffs and many other products, either alone or in conjunction with other inhibitors. Citric acid, for example, is effective at stabilizing soybean oil at a concentration of 0.01%, or 5×10^{-4} M (Madhavi et al., 1995), and similar concentrations have been used in preservative formulations for many other food products. By complexing initiator metal ions, polycarboxylic acids can delay or prevent autooxidative damage of readily oxidized species such as vitamin C (Reiss, 1994).

III. QUENCHING OF SINGLET OXYGEN

A. Physical Quenching

Singlet oxygen, once formed, has a number of potential decay pathways available depending on its surroundings. In the stratosphere, for example, where relatively few other molecules surround it, its lifetime may be on the order of minutes (Wayne, 1991). In solution, however, potential acceptors of its excited-state energy are present at much higher concentrations. Collisions with these molecules can either result in no reaction, or energy transfer from $^1O_2$ to the collision partner. In some cases, a chemical reaction may take place with incorporation of oxygen into the acceptor, but physical or collisional quenching, where $^1O_2$ returns to the ground state and no new chemical compounds are formed, is generally more likely to occur.

Collisional quenching of singlet oxygen occurs with transfer of the excess energy of the excited state to the quencher. The quenching molecule may then dissipate its newly acquired energy by emitting it to the medium as heat. This property may permit the quencher to deactivate many molecules of $^1O_2$.

Quenching by solvents usually follows this type of mechanism. Water, for example, rapidly deactivates $^1O_2$, which has a lifetime of only a few microseconds in aqueous solution; in some organic solvents, however, the lifetime is orders of magnitude greater (Wilkinson and Brummer, 1981).

Solutes must be extremely reactive toward $^1O_2$ to compete with solvent-induced physical quenching. Solvent quenching is a pseudo-first order process, whereas reaction with a solute would be expected to follow second-order kinetics and would therefore have a strong dependence on solute concentration. Again, either chemical reaction or physical quenching may take place. Many studies have shown that the relative quantum yields of chemical reaction versus physical quenching vary according to the solvent environment and the acceptor concentration. In addition, physical quenching can be promoted by acceptors having low-lying triplet states, or by favorable charge-transfer properties of an acceptor $-^1O_2$ complex (Wilkinson and Brummer, 1981; Chignell et al., 1994).

Many synthetic and naturally occurring compounds have been shown to deactivate singlet oxygen primarily by physical quenching (Bellus, 1979). These include amines (including some plant alkaloids), carotenoids, vitamin E, and transition-metal salts and complexes. Some naturally occurring metalloporphyrins such as vitamin B_{12} derivatives with Co(II) centers have been shown to be extremely efficient physical quenchers of $^1O_2$ (Oliveros et al., 1995).

B. Quenching with Reaction

Naturally occurring compounds that are very reactive with $^1O_2$ (see Chapter 2) could in theory act as quenchers in living cells. It would be expected that such compounds would be most effective if they were present at rather high concentrations. A few classes of compounds have particularly high quantum yields of reaction and might act as defense substances. Cyclic dienes such as α-terpinene, furans, and histidine have quantum yields of reaction that are close to 1.0 (Wilkinson and Brummer, 1981).

IV. DESTRUCTION OF PEROXIDES

In Chapter 2 it was shown that a key step for increasing the rate of a radical chain reaction was the accumulation of hydroperoxides, ROOH, that act as branching points to stimulate the production of reactive free radicals. In theory, if the formation of these compounds could be slowed, or if they could be converted to inactive substances once they were formed, the propagation phase could be eliminated or greatly delayed. This has been the rationale for developing "peroxide-destroying" antioxidant additives.

As we have already mentioned (Chapter 2), reducing transition metals are effective reagents that bring about peroxide decomposition, but where these reactions occur by one-electron mechanisms, the result is the formation of a reactive free radical:

$$ROOH + M^{n+} \to RO\cdot + M^{(n+1)}$$

(Equation 3-1)

Reactions of this nature occur both inside and outside living cells. They appear to be involved, for example, in the decomposition of lipid hydroperoxides by iron(II)-containing heme proteins, with the formation of damaging lipid alkoxy radicals (Aoshima et al., 1986). Therefore, the only useful metals for hydroperoxide inactivation would be those that could contribute two electrons.

Most hydroperoxides are not particularly strong oxidizing agents, and therefore only quite readily reduced organic compounds are very effective for reducing their concentrations. Synthetic peroxide destroyers fall into two categories: metal complexes and reduced sulfur and phosphorus compounds (usually phosphites and thioesters: Sousa, 1995). Metallic salts like thiophosphates and thiocarbamates are sometimes used (Carlsson and Wiles, 1974); some of these compounds also play another (and perhaps related) role as UV light stabilizers.

Hydroperoxides may also be decomposed by reducing agents. Again, a one-electron reaction will give rise to a new radical species, probably a far more reactive oxidant than ROOH. Addition of a solvated electron, for example, to t-butyl hydroperoxide affords the t-butoxyl radical (Phulkar et al., 1990):

$$t\text{-BuOOH} + e_{aq}^- \to t\text{-BuO}\cdot + HO^-$$

(Equation 3-2)

This reaction was found to be fast ($k = 5 \times 10^9$ l/mol sec) compared to a much slower reaction with H· (5×10^7 l/mol sec) and the almost nonexistent reaction with ·O_2^- (5 l/mol sec).

In organisms a variety of enzymes act as peroxide-destroying agents. Catalase, for example, decomposes hydrogen peroxide, and peroxidases reduce hydroperoxides. However, many other molecules found in cells also have the property of reacting with peroxides or other cellular oxidants and removing them from autooxidative reactions. Although most of these studies have been done with "peroxidizing lipids," which would contain multiple reactive species, a few have been performed with purified hydroperoxides. For example, several enzymes react *in vitro* with linoleic acid hydroperoxide with extensive loss of the amino acids methionine, tyrosine, histidine, and lysine (Gamage and Matsushita, 1973). Individual amino acids (those listed above as well as cysteine) also react with hydroperoxides and other lipid peroxidation products (Gardner, 1979; Kikugawa et al., 1991). Some of these reactions may be catalyzed by trace amounts of iron or other metal ions (Gardner et al., 1976).

Monomeric sulfur compounds such as thiols and dihydrolipoic acid (3-6), are rather easily oxidized by many electrophiles such as hydroperoxides, hydrogen peroxide, superoxide, singlet oxygen, and hydroxyl radical (Asada and Kanematsu, 1976, Suzuki et al., 1991). Typical products of these reactions are, in the case of thiols, the corresponding disulfides and sulfonic acids; and in the case of sulfides, the corresponding sulfoxides and sulfones.

$$HS-CH_2-CH_2-CH\begin{matrix}(CH_2)_4COOH\\ \\ SH\end{matrix}$$

(3-6)

Although most phenols do not react readily with hydroperoxides, their anions may. In the presence of bases such as pyridine, for example, the reaction rate of p-methoxyphenol with cumene hydroperoxide was accelerated (Martemyanov et al., 1972).

V. SYNERGISTIC MECHANISMS

Synergism is defined in chemical terms as an observed effect on some reaction or process that cannot be explained merely by taking into account the effects of individual participants. In autooxidation reactions where two or more reaction partners are involved, the rate of oxidation could either be faster or slower than expected. In the usual use of the term, it refers to the effect of two or more antioxidant additives which individually are less effective than their combination (Jacob, 1995). For example, a metal-complexing agent such as citric acid might be added to a material together with a radical scavenger like BHT. Other possibilities would include adding two radical scavengers, a light stabilizer and a radical scavenger, or a radical scavenger and a peroxide destroyer. All of these combinations and more have been used for protecting synthetic materials and foods, and all also seem to be involved in cellular defense.

It is often far easier to observe synergism than to explain it. The mechanism of action of a given antioxidant may not be entirely clear-cut; for example, radical inhibitors such as polyphenols may also have metal-complexing properties. In extreme cases, such as EDTA–BHT combinations, the observed effects probably have relatively simple explanations. However, many other combinations of additives are still incompletely understood. Even in a case where the additive is, for example, a good metal-binding agent in simple systems, the interpretation of its actions in more complex environments may be more difficult. The metallic species in the complex system may be present in a variety of forms, some of which may readily be given up to the additives and others not. The result would be a partial, but not a complete, cessation of metal-promoting behavior. Examples of other mechanistic possibilities may readily be envisaged.

Perhaps the best-studied synergistic relationship in living cells is the vitamin E–vitamin C combination (Niki et al., 1982; Barclay et al., 1985). Vitamin E reacts with peroxyl radicals to give the hydroperoxide and an intermediate tocopheryl radical. Subsequent reaction with vitamin C, which is usually present in much higher concentration and therefore acts as a "sacrificial" partner, regenerates vitamin E through hydrogen donation:

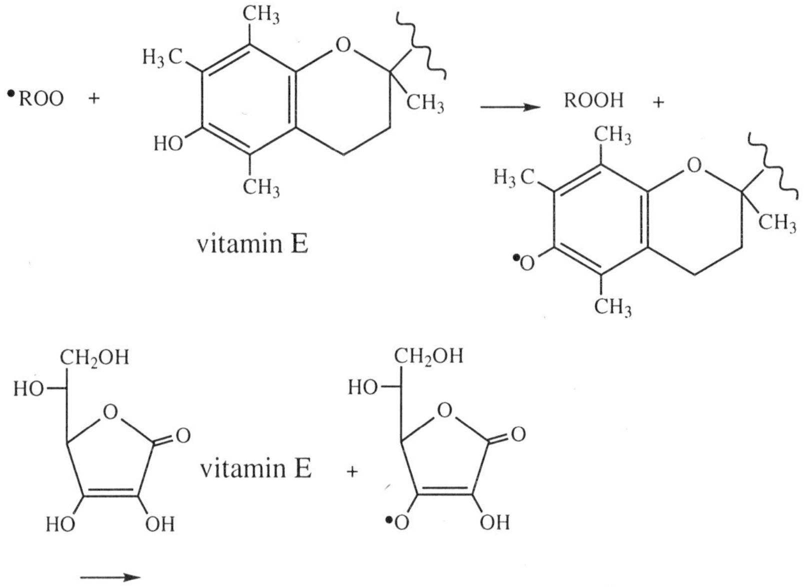

•ROO + vitamin E ⟶ ROOH + [tocopheryl radical]

vitamin E + [ascorbate] ⟶

(Equation 3-3)

This reaction has been demonstrated to have a comparable rate to that of the reaction of vitamin E radicals with peroxyl radicals (Burton and Ingold, 1981). Several other naturally occurring compounds, including some amino acid and indole derivatives, may react with vitamin E in a similar electron-transfer mechanism (Cadenas et al., 1989).

REFERENCES

Alumaa, P. and J. Pentsub. 1994. Prediction of chromatographic retention of metal complexes from stability constants measured by ion chromatography. Chemosphere 38, 566–570.

Aoshima, H., Y. Yoshida, and H. Taniguchi. 1986. Reaction between lipid hydroperoxide and hemoglobin studied by a spectrophotometric and a spin trapping method. Agric. Biol. Chem. 50, 1777–1783.

Barclay, L.R.C., Locke, S.J., and MacNeil, J.M. 1985. Autoxidation in micelles. Synergism of vitamin C with lipid-soluble vitamin E and water-soluble Trolox. Can. J. Chem. 63, 366–374.

Bellus, D. 1979. Physical quenchers of molecular oxygen. Adv. Photochem. 11, 105–205.

Berenbaum, M. R. 1987. Charge of the light brigade: phototoxicity as a defense against insects. In J. R. Heitz and K. R. Downum, eds., *Light-Activated Pesticides.* Amer. Chem. Soc. Sympos. Ser. #339, pp. 206–216. Washington, D.C.

Berthon, G. 1995. Critical evaluation of the stability constants of metal complexes of amino acids with polar side chains. Pure Appl. Chem. 67, 1117–1240.

Buettner, G. R. 1986. Ascorbate autoxidation in the presence of iron and copper chelates. Free Rad. Res. Commun. 1, 349–353.

Buettner, G. R. 1993. The pecking order of free radicals and antioxidants: lipid peroxidation, α-tocopherol, and ascorbate. Arch. Biochem. Biophys. 300, 535–543.

Burton, G.W. and Ingold, K.U., 1981. Autoxidation of biological molecules. 1. The antioxidant activity of vitamin E and related chain-breaking phenolic antioxidants in vitro. J. Am. Chem. Soc., 103, 6472–6477.

Cadenas, E., Simic, M.G., and Sies, H. 1989. Antioxidant activity of 5-hydroxytryptophan, 5-hydroxyindole, and DOPA against microsomal lipid peroxidation and its dependence on vitamin E. Free Rad. Res. Commun., 6, 11–17.

Carlsson, D. J. and D. M. Wiles. 1974. Photostabilization of polypropylene. II. Stabilizers and hydroperoxides. J. Polym. Sci. Polym. Chem. Ed. 12, 2217–2223.

Chignell, C. F., P. Bilski, K. J. Reszka, A. G. Motten, R. H. Sik, and T. A. Dahl. 1994. Spectral and photoehemical properties of curcumin. Photochem. Photobiol. 59, 295–302.

Gamage, P. T. and S. Matsushita. 1973. Interactions of the autoxidized products of linoleic acid with enzyme proteins Agric. Biol. Chem. 37, 1–8.

Gardner, H. W. 1979. Lipid hydroperoxide reactivity with proteins and amino acids: a review. J. Agric. Food Chem. 27, 220–229.

Gardner, H. W., D. Weisleder, and R. Kleiman. 1976. Addition of N-acetylcysteine to linoleic acid hydroperoxide. Lipids 11, 127–134.

Graff, L., G. Muller, and D. Burnel. 1995. *In vitro* and *in vivo* evaluation of potential aluminum chelators. Veterinar. Human Toxicol. 37, 455–461.

Gutteridge, J. M. C., D. A. Rowley, and B. Halliwell. 1982. Superoxide-dependent formation of hydroxyl radicals and lipid peroxidation in the presence of iron salts. Biochem. J. 206, 605–609.

Hogfeldt, E. 1982. Stability constants of metal ion complexes. Part A. *Inorganic Ligands.* Pergamon, Oxford.

Jacob, R. A. 1995. The integrated antioxidant system. Nutr. Res. 15, 753–766.

Jovanovic, S., S. Steenken, and M. G. Simic, 1990. One-electron reduction potentials of 5-indoxyl radicals. A pulse radiolysis and laser photolysis study. J. Phys. Chem. 94, 3583–3588.

Jovanovic, S. J., M. Tosic, and M. G. Simic. 1991. Use of the Hammett correlation and σ^+ for calculation of one-electron redox potentials of antioxidants. J. Phys. Chem. 95, 10824–10827.

Kikugawa, K., T. Kato, and A. Hayasaka. 1991. Formation of dityrosine and other fluorescent amino acids by reaction of amino acids with lipid hydroperoxides. Lipids 26, 922–929.

Krinsky, N. I. 1992. Mechanism of action of biological antioxidants. Proc. Soc. Exp. Biol. Med. 200, 248–254.

Lind, J., X. Shen, T. E. Eriksen, and G. Merenyi. 1990. The one-electron reduction potential of 4-substituted phenoxyl radicals in water. J. Am. Chem. Soc. 112, 479–482.

Madhavi, D. L., R. S. Singhal, and P. R. Kulkarni. 1995. Technological aspects of food antioxidants. In Madhavi, D. L., S. S. Deshpande, and D. K. Salunkhe, eds. *Food Antioxidants: Technological, Toxicological, and Health Perspectives.* Marcel Dekker, New York, pp. 159–265.

Martemyanov, V. S., E. T. Denisov, and V. V. Fedorova. 1972. Bases as catalysts for the oxidation of p-methoxyphenol by cumyl hydroperoxide. Kinet. Katal. 13, 303–307.

Meites, L., ed. 1963. *Handbook of Analytical Chemistry.* McGraw-Hill, New York.

Moore, W. J. 1962. *Physical Chemistry,* 3rd ed. Prentice-Hall, New York.

Niki, E., Tsuchiya, J., Tanamura, R., and Kamiya, Y., 1982. Oxidation of lipids. II. Rate of inhibition of oxidation by α-tocopherol and hindered phenols measured by chemiluminescence. Bull Chem. Soc. Japan 55, 1551–1555.

Odani, A. and O. Yamauchi. 1996. Stability constants of metal compounds of amino acids with charged side chains. 1. Positively charged side chains. Pure Appl. Chem. 68, 469–496.

Oliveros, E., F. Besancon, M. Boniva, B. Krauther, and A. M. Braun. 1995. Singlet oxygen ($^1\Delta_g$) sensitization and quenching by vitamin B-12 derivatives. J. Photochem. Photobiol. B29, 37–44.

Oturan, M., P. Dostert, M. S. Benedetti, J. Moiroux, A. Anne, and M. B. Fleury. 1988. One-electron and two-electron reductions of 1-methyl-4-phenylpyridinium (MPP+). J. Electroanal. Chem. Interfacial Electrochem. 242, 171–179.

Page, M. I. and A. Williams, eds. 1987. *Enzyme Mechanisms.* Royal Society of Chemistry, London.

Pankow, J. F. 1991. *Aquatic Chemistry Concepts.* CRC Press/Lewis Publishers, Boca Raton, FL.

Perrin, D. D. and L. G. Sillen. 1979. Stability constants of metal ion complexes. Part B. *Organic Ligands.* Pergamon, Oxford.

Phulkar,S., B. S. M. Rao, H.-P. Schuchmann, and C. von Sonntag. 1990. Radiolysis of tertiary butyl hydroperoxide in aqueous solution. Reductive cleavage by the solvated electron, the hydrogen atom, and, in particular, the superoxide radical anion. Z. Naturforsch. 45b, 1425–1432.

Rauser, W. E. 1995. Phytochelatins and related peptides: structure, biosynthesis, and function. Plant Physiol. 109, 1141–1149.

Reiss, J. 1994. Stability of ascorbic acid (Vitamin C) in tea. Ernahr. Umschau 41, 342 (only).

Schwarzenbach, R. P., P. M. Gschwend, and D. M. Imboden. 1993. *Environmental Organic Chemistry.* John Wiley, New York, NY.

Sousa, M. 1995. Antioxidants. Mod. Plastics, mid-November, pp. C4–C5.

Stanbury, D. M. 1989. Reduction potentials involving inorganic free radicals in aqueous solution. Adv. Inorg. Chem. 33, 69–138.
Stumm, W. and J. J. Morgan. 1981. *Aquatic Chemistry,* 2nd ed. Wiley-Interscience, New York.
Wayne, R. P. 1991. *Chemistry of Atmospheres.* 2nd ed. Clarendon Press, Oxford.
Yamauchi, O. 1995. Amino acid metal and pterin metal chemistry as an approach to biological functions. Pure Appl. Chem. 67, 297–304.

4 KINETICS AND MECHANISMS OF INHIBITED AUTOOXIDATION

I. KINETICS OF AUTOOXIDATION IN THE ABSENCE OF INHIBITORS

A general and simplified mechanism for initiated autooxidation, as discussed in Chapter 2, may be given in outline form, together with the rate constants for each reaction in the sequence:

Initiation: $$RH + In \xrightarrow{k_i} R\cdot + \text{In-derived products}$$

(Equation 4-1)

Propagation: $$R\cdot + O_2 \xrightarrow{k_{ox}} ROO\cdot$$

(Equation 4-2)

$$ROO\cdot + R'H \xrightarrow{k_p} ROOH + R'\cdot$$

(Equation 4-3)

Termination: $$2\,ROO\cdot \xrightarrow{k_t} \text{nonradical products}$$

(Equation 4-4)

A kinetic analysis for the simplified situation will be presented first, followed by a discussion of how the presence of antioxidants affects the observed results.

A. Initiation

In the early stages of an autooxidation, the rate of the process is dominated by the initiation step. In most instances, this period features a relatively long interval during which few chemical changes take place in the system; in fact, the phase is often called the "lag period." The rate of the reaction during the lag period is a consequence of the concentration of the initiating agent, In, and the rate constant for its reaction with the substrate RH. In kinetic terms, the rate of initiation (R_i) is the product of these two quantitities:

$$R_i = k_i[In]$$

(Equation 4-5)

In practice, R_i is often determined by measuring the changes in concentration of some reactant that is not the initiator species itself, such as oxygen, for reasons described more fully later. The determination of k_i can be difficult or impossible in "spontaneous" or cellular oxidation processes where the identities and concentrations of possible initiating agents is not clear. However, in deliberately promoted autooxidations, where a kinetically well-behaved initiator is added, k_i can readily be derived. Most careful studies have employed an azo initiator of the general formula R–N=N–R, which (as mentioned in Chapter 2), although quite stable at ordinary room temperature or below, decomposes thermally to molecular nitrogen and alkyl free radicals (Boozer et al., 1955; also see Barclay et al., 1984 and references therein). Several of these compounds have been widely used, including azo-*bis*-isobutyronitrile (AIBN: 4-1a), which readily breaks down to 2-cyano-2-propyl radicals at temperatures in the neighborhood of 40–60°C (Walling and Huyser, 1963). Since AIBN is relatively insoluble in water, many studies have been conducted using more polar analogs such as azo-*bis*-4-cyanovaleric acid (ACVA: 4-1b) The loss rate of the initiator can be determined by observing the decrease in its concentration using high-performance liquid chromatography. Using these techniques, it can be seen that a small fraction of initiator radical pairs recombine (probably within the solvent cage in which the thermolytic decomposition first took place). The fraction of radical recombination depends strongly on the solvent employed. Normally, however, the loss rate is a fairly straightforward first-order process (Blackley and Haynes, 1979) with ACVA, for example, having a disappearance rate constant in pure water of 1.83×10^{-6}/sec at 50°C,

$$(H_3C)_2\underset{\underset{CN}{|}}{C}-N=N-\underset{\underset{CN}{|}}{C}(CH_3)_2$$

(4-1a)

$$\text{HOOC}-CH_2-CH_2-\underset{\underset{CN}{|}}{\overset{\overset{CH_3}{|}}{C}}-N=N-\underset{\underset{CN}{|}}{\overset{\overset{CH_3}{|}}{C}}-CH_2-CH_2-\text{COOH}$$

(4-1b)

corresponding to a half-life of about 100 hr. Since the initial concentration of the initiator is known, R_i can readily be determined by Equation 4-5, or by a surrogate method.

Because azo initiators produce two equivalents of radicals per mole, a kinetic factor of ½ needs to be entered into rate expressions using them. That is, the oxidation rate should be one-half-order in initiator concentration (or should depend on the square root of the initiator concentration). Other classes of initiators that undergo homolysis to give two moles of radicals (such as peroxides) also show this dependence.

In theory, an initiator never decomposes quantitatively to free radicals, and the kinetic expression involving the rate of initiation should be corrected by a factor, f, that represents the fraction of conversion of the initiator. In practice, however, commonly used thermal initiator compounds have $f \approx 1.0$. However, in situations where the initiator is unknown, or is generated by some outside influence such as UV illumination, f may need to be determined by some independent technique.

B. Propagation

The chain process continues with the reactions of the initially formed alkyl radicals with oxygen to form alkylperoxyl radicals and the subsequent chain-carrying reactions of the peroxyl radicals. Ordinarily, the rate constants k_{ox} are neglected in coming up with an overall rate expression, since, with few exceptions, they are so large as to approximate diffusion control ($k \approx 2 \times 10^9$ l/mol sec). However, the ROO· radicals react much more slowly with most hydrogen donors, making k_p an important controlling parameter for many autooxidation reactions.

Propagation rate constants have been determined by a variety of means, such as the rotating-sector method, chemiluminescence techniques, electron spin resonance, pulse radiolysis, etc. Second-order rate constants for the abstraction of a primary alkane hydrogen atom by ROO· are approximately 10^{-5} l/mol sec per hydrogen; for secondary hydrogens, the rate increases by tenfold or somewhat more; and for a tertiary hydrogen another factor of 10 comes into play (Hendry et al., 1974). In general, a reasonable inverse correlation exists between log k for H-abstraction and C–H bond strengths. Cyclic alkanes are somewhat more reactive, with rate constants in the 0.1–1.0 l/mol sec range (Denisov, 1995).

For alkenes, reactivity is normally higher, especially if the double bond is substituted by electron-donating substituents such as methyl groups, if it is part of a conjugated system, or if it is contained in a ring (either alicyclic or aromatic). Second-order rate constants for ROO·-unconjugated alkene reactions are usually within an order of magnitude of 2 l/mol sec (Howard, 1984). With this class of compounds, abstraction of an allylic hydrogen normally predominates, but a competing reaction, addition to the double bond, may also occur, as discussed in Chapter 2:

$$ROO\bullet + >C=C< \rightarrow ROO-\overset{|}{\underset{|}{C}}-\overset{\bullet}{C}<$$

(Equation 4-6)

Electronic and steric factors are very important in the further fates of these radicals. Collapse to an epoxide and alkoxy radical can occur if the oxygen concentration is small, but since the latter reaction is almost diffusion-controlled, the first-order epoxide formation rate constant must be large in order for it to be competitive (Bloodworth et al., 1984).

In the initiation of lipid peroxidation, a peroxyl radical tends to abstract a hydrogen from a bis-allylic methylene position if one is available, as it is, for example, in linoleic or linolenic acid derivatives. The second-order rate constants for the reaction vary to some degree with solvent, but are typically rather higher than for most olefins (k = 30–60 l/mol sec: Howard and Ingold, 1967; Barclay et al., 1990). Since lipids have high local concentration in membranes, the rate of the reaction can be significant.

Benzylic H-atoms are also somewhat more susceptible to ROO· attack than ordinary C–H hydrogens, having rate constants in the range of 0.1–3 l/mol sec (Hendry et al., 1974; Denisov, 1995). Again, tertiary or ring-contained benzylic hydrogens have the higher rate constants in the series. These factors are important in the autooxidation of petroleum products, particularly fuel oils, which are rich in compounds with tertiary benzylic groups such as partly hydrogenated and alkyl-substituted indenes and naphthalenes.

For nonhydrocarbon compounds, the introduction of an oxygen atom to form aliphatic alcohols or ethers has little effect on the kinetics. Aldehydes, however, possess a labile hydrogen and have rate contants for H-abstraction of roughly 1 l/mol sec (Korcek et al., 1972). Incorporation of phenolic or heterocyclic O, however, may cause profound enhancements in peroxyl radical rate constants.

Many phenols are active inhibitors of oxidation, of course, and this activity correlates well with their ability to quench peroxyl radicals. A value of about 10^3–10^5 l/mol sec is a good approximation for the rate constant for the reaction of many simple phenols with ROO· (Howard, 1972; Howard and Furimsky,

1973; Pohlman and Mill, 1983; Denisov, 1995). Therefore, even a very small concentration of a phenolic antioxidant can inhibit the autooxidation of a much larger concentration of a less reactive substrate. The reaction appears to involve the specific extraction of the hydrogen of the phenolic OH, giving a free radical with odd-electron character not only at the oxygen atom, but also at o and p positions of the aromatic ring. Phenols also react (although rather slowly; second-order rate constants on the order of $\sim 10^{-3}$ l/mol sec) with hydroperoxides, but in this case the product is a new RO· radical that can initiate new chains (Martemyanov et al., 1972).

For an antioxidant to compete, the rate of its reaction (the product of its rate constant and local concentration) must be comparable or greater to the rate of the initiation step. For the case of vitamin E, its second-order rate constant with peroxyl radicals is about 10^6 l/mol sec in solution (though it may be orders of magnitude less in some lipid bilayers: cf. Barclay et al., 1990), and therefore should be able to inhibit peroxidation if its concentration is more than about 1/20,000 of that of a peroxidizable lipid with a second-order rate constant of 50 l/mol sec. The kinetics of reactions involving all these compounds will be discussed further in the next section, and other characteristics of the individual phenolic antioxidants are discussed in Chapter 5.

Similar rate constants, or even larger ones, are also observed for the reaction of many aromatic amine derivatives with ROO· (Howard and Furimsky, 1973; von Sonntag, 1987), although in this case it is often electron transfer to the peroxyl radical, rather than hydrogen atom transfer, that predominates:

$$ArNH_2 + ROO· \rightarrow ArNH_2^{+}· + ROO^-$$

(Equation 4-7)

Other heterocyclic compounds may also be quite reactive with ROO·, in keeping with the use of many such compounds as antioxidants. Ascorbate (vitamin C), for example, reacts with peroxyl radicals of various structural characteristics with rate constants in the 10^4–10^8 range (Packer et al., 1980; Ariga and Hamano, 1990). Further discussions of the chemistry of ascorbate will be found in Chapter 5.

C. Termination

In the final stages of a chain reaction, when all the oxygen or active hydrogen species are used up, the termination phase begins. In this phase, the radicals recombine with each other to produce inactive, nonradical products (reaction 4-4), a process that (for peroxyl radicals, at least) is usually slower than typical radical-radical coupling reactions. The products of these reactions were discussed in Chapter 2. The recombinations normally eliminate two equivalents of chain-carrying radicals, and accordingly a factor of 2 needs to be incorporated into the rate expression.

Rate constants for chain termination in alkane autooxidations depend strongly on structure, and usually fall in the order primary > secondary > tertiary (Howard and Ingold, 1967; Reich and Stivala, 1969). There is a fair degree of variability among structurally related radicals, probably reflecting differences in steric and polar effects. For primary peroxyl radicals, k_t is often close to 10^8 l/mol sec, secondary radicals about 10^7, and tertiary around 10^5.

D. Overall Rate Law

Under the conditions discussed above, the rate law for oxygen uptake by an autooxidizing system containing a single compound, with oxygen in excess, is given by Equation 4-8.

$$-\frac{d[O_2]}{dt} = \frac{[RH]k_p(R_i)^{1/2}}{(2k_t)^{1/2}}$$

(Equation 4-8)

In the expression, the rate of oxidation is described in terms of oxygen uptake. Many kinetic studies of autooxidation have used this property, but in systems where the concentration of the material being oxidized is small, the measurement can be quite difficult or may require specialized equipment. Often, another measure of autooxidation, such as the appearance of some major product, is used as a surrogate parameter for the rate.

When both the rate of oxygen uptake and the initiation rate are known, a quantity called the kinetic chain length (KCL) can be determined. This dimensionless parameter is equivalent to the number of substrate molecules consumed per molecule of initiating species, and is a measure of the efficiency of chain propagation:

$$KCL = (-d[O_2]/dt)/R_i$$

(Equation 4-9)

The kinetics of peroxidation of lipids (and, indeed, of similar hydrocarbon-like autooxidative chain reactions) display a complex dependence on initiator, substrate, and oxygen concentrations, as well as temperature. Variabilities in the rate constants for initiation, propagation, and termination steps can result in deviations from the "typical" rate law of Equation 4-8. In one study of polyunsaturated fatty acids, in which the effects of initiator, lipid, and oxygen concentration were systematically examined, the reaction displayed a half-order dependence on each of the three parameters. That is, when the concentration of one partner was increased fourfold while the other two were kept constant, the overall rate of peroxidation doubled (Dirks et al., 1982).

The authors interpreted this result to mean that in these particular reactions, the steady-state concentrations of R· and ROO· were comparable and that a cross-termination reaction,

$$R\cdot + ROO\cdot \rightarrow ROOR$$

(Equation 4-10)

predominated over the more typical ROO· dimerization process, Equation 4-4. Other workers, however, have not found this relationship to hold and consider peroxidation of lipids to closely follow the more traditional rate law (Cosgrove et al., 1987).

II. AUTOOXIDATION IN THE PRESENCE OF INHIBITORS

Autooxidation kinetics are drastically changed upon the introduction of compounds that can interfere with either its initiation or propagation. Additives to autoxidizing mixtures are sometimes divided into *inhibitors* (true antioxidants, those that effectively stop radical processes until they are consumed) and *retarders* (compounds that merely slow autooxidation). The kinetic effects of the two classes on autooxidation are not simply summed up, but one type is schematically illustrated in Figure 4-1. Uninhibited autooxidation is represented by curve A, with oxygen uptake, peroxide formation, or some other measure of oxidation plotted on the Y-axis. A retarder (curve B) may decrease the autooxidation rate over the entire period of observation. A true inhibitor, however, induces a prolonged lag period (curve C), and when the lag period ends autooxidation commences with a rate that is essentially the same as that for the uninhibited reaction.

Antioxidants act by different mechanisms and, especially in biochemistry, numerous substances may be present that can effectively minimize autooxidation by interfering with numerous steps in the initiation and propagation pathway. A detailed examination of inhibited autooxidation in nonbiological systems has been presented by Denisov (1995) in a volume containing many useful tables of rate constants and other relevant data.

A. Inhibition by Preventive Antioxidants

1. *Complexing Agents*

An inhibitor may act by quenching, or reducing the concentration of, either initiating or propagating species. Agents such as chelators, which bind metal ions into an inactive form, for example, may prevent an oxidation from beginning at all, slow it by reducing the concentration of active initiators, or prevent it from continuing (if metal-catalyzed decomposition of H_2O_2 or hydroperoxides is an important propagation step). Within the cell, this inhibitory

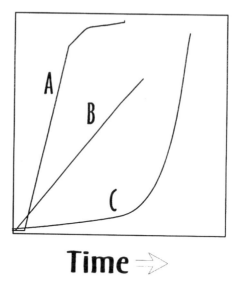

Figure 4-1 Course of autooxidation in the presence and absence of inhibitors. Vertical axis represents the magnitude of a property such as oxygen uptake that occurs concomitantly with substrate oxidation. Curve A = course of autooxidation in the absence of inhibitor. Curve B = course of autooxidation in the presence of a retarder. Curve C = course of reaction in the presence of a chain-breaking antioxidant. For details, see text.

mechanism is likely to be extremely important since so many biologically important processes of oxidative damage are promoted by "free" iron ions.

Numerous biomolecules have the potential to complex metallic species, including citric acid and other polycarboxylic acids, hydroxy acids, peptides, and amino acids. In order to assess the importance of complexation on the rate of autooxidation, it is necessary to be able to calculate the concentrations of active metallic initiator species in the presence of potential complexing agents. Obviously this would be a formidable task in a living cell. Therefore, at present we are limited to *in vivo* estimates. Some of the methods used for the calculation of stability constants for metal complexes were briefly discussed in Chapter 2. Advantage is taken of potent complexing agents in chelation therapy (Halliwell, 1989), where deferoxamine, a naturally occurring hydroxamate derivative, and similar compounds that strongly bind iron are employed to manage acute and chronic disease situations (e.g., anemia, ischemia-reperfusion injury, and hemorrhage).

2. Light Absorbers

Light-induced autooxidations may also be stopped or slowed by various means. For example, photons can be reflected from a surface, or trapped by using a strongly opaque shielding substance, as exemplified by the addition of carbon black to rubber to prevent the penetration of light into the product.

Similarly, free-radical scavengers such as β-carotene, when present in photosynthetic tissues exposed to light, may react so rapidly with triplet excited states that they are precluded from attacking biomolecules such as proteins. The process reduces the rate of chain initiation in the cell and is therefore an example of preventive antioxidation.

3. Peroxide Decomposers

Compounds that react with peroxyl radicals, a one-electron oxidation, are generally assumed to be "chain-breaking" antioxidants and are discussed in the following section. Inhibition of the propagation step may also take place by agents that chemically reduce or otherwise decompose hydroperoxides, ROOH, by a two-electron process. Since hydroperoxide decomposition during a chain reaction usually contributes to the acceleration of autooxidation, their removal by additives may significantly slow the process. Many antioxidants are assumed to act, at least partially, by this mechanism. Synthetic antioxidants containing reduced sulfur and phosphorus functional groups have been shown to be effective peroxide decomposers, as have a variety of nickel- and zinc-containing complexes having thiophosphate and thiocarbamate ligands (Scott, 1990).

In biological environments thiols and sulfides, RSH and RSR, are readily oxidized by both one- and two-electron reactions, and theoretically could be involved in both types of inhibition. The principal pathway followed would be a function of the relative concentrations of ROO· and ROOH at any point in the oxidation process, as well as the relative rate constants of the two oxidants toward the sulfur compound.

Certain other sulfur compounds do not appear to interact readily with chain-carrying radicals, but do reduce hydroperoxides to alcohols, apparently by a two-electron, nonradical process. Sulfoxides (RS=O) are particularly effective in this mode (Bateman et al., 1962). It has been postulated that the formation of a molecular, and possibly hydrogen-bonded, complex between the hydroperoxide and the sulfoxide is necessary; the suggestion is supported by the requirement for roughly stoichiometric amounts of the sulfoxide (relative to [ROOH]) for efficient inhibition.

B. General Free Radical Scavengers

In theory it should be possible to inhibit radical-initiated autooxidation with some "general" scavenger, perhaps a substance that was a stable free radical in its own right and, after reaction with another radical, was converted to even-electron forms. At one time diphenylpicrylhydrazyl (4-2) was thought to be such an "ideal" reagent (Walling, 1957) on the basis of its efficient inhibition of vinyl polymerization. However, further investigation of its reactions have shown that some are reversible, and others proceed via charge-transfer intermediates that have their own radical chemistry. Therefore, subsequent research has been directed toward identifying compounds that are more selective inhibitors of particular types of radicals.

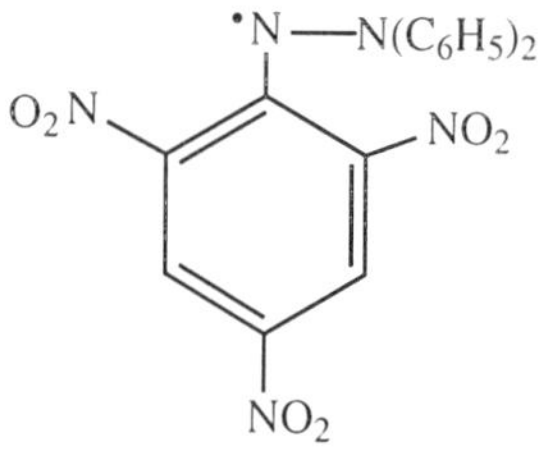

(4-2)

C. Inhibition by Alkyl Radical Scavengers

A special case of preventive autooxidation exists in environments where the oxygen concentration is low. In these situations, the radicals that are formed in the initiation step (Reaction 4-1) may be intercepted and deactivated by other species. These reactions could be important in many types of living cells, where the oxygen concentration may be well below that in the atmosphere.

Chain termination by compounds that react readily with alkyl radicals is well known. Such compounds include quinones and quinone methides, nitro compounds, and stable free radicals.

There are several possible mechanisms for quinone-alkyl radical reactions. Addition of the radical to the C=C double bond is one; attack at the C=O double bond, with the formation of a new ether linkage, is a second; and hydrogen atom donation from the radical, with the formation of an unreactive semiquinone, is a third (Figure 4-2).

Nitro compounds are generally somewhat less reactive with alkyl radicals. Their mechanism of action is not completely established; aromatic compunds, which have been the most frequently studied, may form addition or substitution products with the radical at either the ring or nitro group.

D. Inhibition by Chain-Breaking Antioxidants

1. Competition

In the presence of an inhibitor AH that reacts principally (as many do) with ROO· radicals, the kinetics of inhibition reduce to measuring the competition for these radicals with the species R'H (in Equation 4-2) and AH:

$$ROO \cdot + AH \xrightarrow{k_{inh}} ROOH + A \cdot$$

(Equation 4-11)

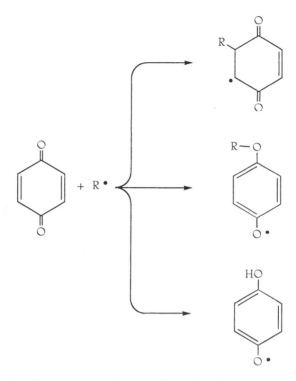

Figure 4-2 Possible reactions of quinones with alkyl radicals.

The assumption is made that the inhibitor is not so readily oxidized as to react directly with molecular oxygen:

$$AH + O_2 \rightarrow A\cdot + HOO\cdot$$

(Equation 4-12)

This presumption is usually justifiable at ordinary temperatures, although in a few cases amines seem to undergo the reaction (Elley, 1936), and electron transfer from O_2 to many complexed transition metal cations does, of course, occur readily.

It is further assumed in this treatment that $A\cdot$ is not an effective chain-carrying radical, and therefore that the chain reaction's propagation stage is slowed down. In cases where $k_{inh} \gg k_t$, Equation 4-8 can be replaced by

$$-\frac{d[O_2]}{dt} = \frac{[RH](R_i)^{1/2} k_p}{k_{inh}[AH]}$$

(Equation 4-13)

In some cases, however, notably with phenolic antioxidants, the inhibitor reacts with two moles of peroxyl radicals. This is a consequence of the reaction of ROO· with the antioxidant to give a free radical A· capable of further reaction:

$$ROO\cdot + AH \rightarrow ROOH + A\cdot \xrightarrow{ROO\cdot} A-OOR$$

(Equation 4-14)

In this case, the quantity k_{inh} in the rate expression is replaced by 2 k_{inh}. The stoichiometry of inhibition can be determined experimentally in unknown cases by measuring the length of the induction period where chain reactions are being initiated at a known rate, e.g. with AIBN.

The most reliable way of measuring k_{inh} is the so-called "lag period method," in which the characteristic delay in the onset of a standard radical chain reaction is measured in the presence and absence of antioxidant. The lag period is directly proportional to k_{inh} and the concentration of the inhibitor (Boozer et al., 1955). In practice, the calculation is done by plotting the oxygen consumption (or some other measure of oxidation) during the lag period versus the natural logarithm of the quantity $1 - (t/\tau)$, where t is the reaction time and τ the total length of the lag period in seconds (Burton and Ingold, 1981). The slope of this plot is the quantity $k_p \times [RH]/k_{inh}$.

Inhibitor rate constants (usually, their rates of reaction with ROO· generated by azo initiators) have been determined for many phenolic compounds. Although absolute values of these rate constants vary depending on the solvent for the reaction and the radical with which they are reacting, the relative constants vary, more or less predictably, over several orders of magnitude. For example, Uri (1961) tabulated values showing that phenol itself, or phenols bearing an electron-withdrawing group such as nitro or carboxyl, had k_{inh} around 10^3 l/mol sec, but that alkyl substitution increased the rate constant by about an order of magnitude. Others have found that polyphenols such as catechol, hydroquinone, and pyrogallol have k_{inh} of about 10^5, and the best sterically hindered phenolic antioxidants such as BHT and vitamin E can be even higher, in the 10^6–10^8 range (Hunter and Simic, 1983).

2. Diffusion

Porter (1993) has pointed out that some lipid antioxidants are much more effective in media having high surface-volume ratios, such as membranes or lipid droplets, than in bulk media such as packaged butter and other low surface-volume ratio fats and oils. In nature, the former condition predominates. In such an environment, the antioxidant molecule needs to diffuse only a short distance to the lipid-aqueous phase interface, where it can be removed, react further, or be replaced with inwardly-diffusing species. In bulk oils, it has sometimes been observed that antioxidant effectiveness in a series of compounds is inversely correlated with hydrophobicity (Sherwin, 1976), whereas the reverse is often true in dispersed lipid preparations such as liposomes (Porter, 1993).

As a further example of the effect, it has been noted that vitamin E, which is normally an extremely effective inhibitor of oxidation, loses its effectiveness in certain situations such as high concentration levels and becomes a pro-oxidant (Cillard and Cillard, 1980). A probable explanation of the paradox is the observation by Ingold et al. (1993) that vitamin E, and presumably other oxidants that form free radical intermediates upon reaction with peroxyl radicals, have some chain-carrying activity. They need to be regenerated by an external reducing agent, such as ascorbate, to retain their effectiveness. Presumably, this can only occur in lipid phases of a relatively small size, assuming that the other reducing agent is located in the aqueous phase. Therefore, highly hydrophobic antioxidants may need to function in a medium where they can readily diffuse to an interface to be regenerated, and if the interface does not exist or is too far away the potential antioxidant may be rendered ineffectual.

A pictorial representation of this phenomenon has been presented by Buettner (1993), who envisions lipid oxidation taking place within the lipophilic regions of membranes (Figure 4-3), giving rise to a peroxyl radical that, being polar, will tend to migrate to the membrane-cytoplasm interface of the cell or organelle surface. At this interface, it will presumably tend to be in closer proximity to the phenolic, hydrophobic head-group of vitamin E, and therefore ought to undergo unusually favorable hydrogen atom transfer and prevent chain propagation. Since the phenolate radical of vitamin E is located at the interface, it should be accessible to the water-soluble ascorbate anion, which in its turn is an effective hydrogen atom donor to repair the vitamin E radical. The final steps of the inhibitory mechanism are the recycling of the ascorbate radical by enzyme systems and the repair of the damaged lipid by various enzyme systems.

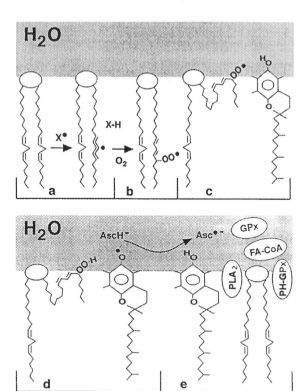

Figure 4-3 Membrane lipid peroxidation. Step a, initiation; b, reaction with molecular oxygen to produce a lipid peroxyl radical; c, partitioning of the peroxyl radical fragment to the membrane-water interface; d, repair of the peroxyl radical by vitamin E and formation of the tocoperoxyl radical; e, repair of the tocopheroxyl radical by ascorbate at the membrane-water interface. (From Buettner, G. 1993. Arch. Biochem. Biophys. 300, 540. Reprinted by permission.)

REFERENCES

Ariga, T. and M. Hamano. 1990. Radical scavenging action and its mode in procyanidins B-1 and B-3 from azuki beans to peroxyl radicals. Agric. Biol. Chem., 54, 2499–2504.

Barclay, L. R. C., S. J. Locke, J. M. MacNeil, J. Van Kessel, G. W. Burton, and K. U. Ingold. 1984. Autoxidation of micelles and model membranes. Quantitative kinetic measurements can be made by using either water-soluble or lipid-soluble initiators with water-soluble or lipid-soluble chain-breaking antioxidants. J. Am. Chem. Soc., 106, 2479–2481.

Barclay, L. R. C., K. A. Baskin, K. A. Dakin, S. J. Locke, and M. R. Vinqvist. 1990. The antioxidant activities of phenolic antioxidants in free radical peroxidation of phospholipid membranes. Can. J. Chem. 68, 2258–2269.

Bateman, L., M. E. Cain, T. Colclough, and J. I. Cunneen. 1962. The antioxidant action of sulfoxides and thiolsulfinates in autoxidizing squalene. J. Chem. Soc. 3570–3578.

Blackley, D. C. and A. C. Haynes. 1979. Kinetics of thermal decomposition of 4,4'-azobis-(4-cyanopentanoic acid) and its salts in aqueous solution. J. Chem. Soc. Faraday Trans. 75, 935–941.

Bloodworth, A. J., J. L. Courtneidge, and A. G. Davies. 1984. Rate constants for the formation of oxiranes by γ-scission in secondary β-t-butylperoxyalkyl radicals. J. Chem. Soc. Perkin Trans. II, 523–527.

Boozer, C. E., G. S. Hammond, C. E. Hamilton, and J. N. Sen. 1955. Air oxidation of hydrocarbons. II. The stoichiometry and fate of inhibitors in benzene and chlorobenzene. J. Am. Chem. Soc., 77, 3233–3237.

Buettner, G. 1993. The pecking order of free radicals and antioxidants: lipid peroxidation, α-tocopherol, and ascorbate. Arch. Biochem. Biophys. 300, 535–543.

Burton, G. W. and K. U. Ingold. 1981. Autoxidation of biological molecules. 1. The antioxidant activity of vitamin E and related chain-breaking phenolic antioxidants *in vitro*. J. Am. Chem. Soc. 103, 6472–6477.

Cillard, J. and P. Cillard. 1980. Behavior of α-, β-, γ- and δ-tocopherols with linoleic acid in aqueous media. J. Am. Oil Chem. Soc. 57, 39–42.

Cosgrove, J. P., D. F. Church, and W. A. Pryor. 1987. The kinetics of the autoxidation of polyunsaturated fatty acids. Lipids 22, 299–304.

Denisov, E. 1995. *Handbook of Antioxidants: Bond Dissociation Energies, Rate Constants, Activation Energies and Enthalpies of Reactions.* CRC Press, Boca Raton, FL.

Dirks, R. C., M. D. Faiman, and E. S. Huyser. 1982. The role of lipid, free radical initiator, and oxygen on the kinetics of lipid peroxidation. Toxicol. Appl. Pharmacol. 63, 21–28.

Elley, H. W. 1936. The protection of rubber and gasoline by antioxidants. Trans. Electrochem. Soc. 69, 239–266.

Halliwell, B. 1989. Protection against tissue damage *in vivo* by desferrioxamine: what is its mechanism of action? Free Radical Biol. Med. 7, 645–651.

Hendry, D. G., T. Mill, L. Piskiewicz, J. A. Howard, and H. K. Eigemann. 1974. A critical review of H-atom transfer in the liquid phase: chlorine atom, alkyl, trichloromethyl, alkoxy, and alkylperoxy radicals. J. Phys. Chem. Ref. Data 3, 937–978.

Howard, J. A. 1972. Absolute rate constants for reactions of oxyl radicals. Advan. Free Rad. Chem. 4, 49–107.

Howard, J. A. 1984. Measurement of absolute propagation and termination rate constants for alkylperoxyls in solution by the hydroperoxide method. Isr. J. Chem. 24, 33–37.

Howard, J. A. and E. Furimsky. 1973. Arrhenius parameters for reactions of *tert*-butylperoxy radicals with some hindered phenols and amines. Can. J. Chem. 51, 3738–3745.

Howard, J. A. and K. U. Ingold. 1967. Absolute rate constants for hydrocarbon autoxidation. VI. Alkyl aromatic and olefinic hydrocarbons. Can. J. Chem. 45, 793–802.

Hunter, E. P. L. and M. G. Simic. 1983. Kinetics of peroxyl radicals with antioxidants. In G. Cohen and R. A. Greenwald, eds., *Oxy Radicals and Their Scavenger Systems.* Elsevier, New York. Vol. 1, pp. 32–37.

Ingold, K. U., V. W. Bowry, R. Stocker, and C. Walling. 1993. Autoxidation of lipids and antioxidation by α-tocopherol and uniquinol in homogeneous solution and in aqueous dispersions of lipids: unrecognized consequences of lipid particle size as exemplified by oxidation of human low density lipoprotein. Proc. Nat. Acad. Sci. U.S.A. 90, 45–49.

Korcek, S., J. H. B. Chenier, J. A. Howard, and K. U. Ingold. 1972. Absolute rate constants for hydrocarbon autoxidation. 21. Activation energies for propagation and the correlation of propagation rate constants with carbon-hydrogen bond strengths. Can. J. Chem. 50, 2285–2297.

Martemyanov, V. S., E. T. Denisov, and L. A. Samoilova. 1972. Reaction of phenols with cumyl hydroperoxide. Izv. Akad. Nauk SSSR, Ser. Khim. 1039–1042.

Packer, J. E., R. L. Willson, D. Bahnemann, and K.-D. Asmus. 1980. Electron transfer reactions of halogenated aliphatic peroxyl radicals: measurement of absolute rate constants by pulse radiolysis. J. Chem. Soc, Perkin Trans. II, 296–299.

Pohlman, A. and T. Mill. 1983. Free radical oxidations in water: decomposition of azoinitiators and oxidation of p-cresol and p-isopropylphenol. J. Org. Chem. 48, 2133–2138.

Porter, W. L. 1993. Paradoxical behavior of antioxidants in food and biological systems. Toxicol. Indust. Health 9, 93–122.

Reich, L. and S. Stivala. 1969. *Autoxidation of Hydrocarbons and Polyolefins: Kinetics and Mechanisms.* Marcel Dekker, New York.

Scott, G., ed. 1990. *Mechanism of Polymer Degradation and Stabilization.* Elsevier, London.

Sherwin, E. R. 1976. Antioxidants for vegetable oils. J. Am. Oil Chem. Soc. 53, 430–436.

Uri, N. 1961. Mechanism of antioxidation. In W. O. Lundberg, ed., *Autoxidation and Antioxidants.* Interscience-Wiley, New York, Vol. 1, pp. 133–169.

von Sonntag, C. 1987. *The Chemical Basis of Radiation Biology.* Taylor and Francis, London.

Walling, C. 1957. *Free Radicals in Solution.* John Wiley & Sons, New York.

Walling, C., and E. S. Huyser. 1963. Free radical additions to olefins to form carbon–carbon bonds. Org. Reactions 13, 91–149.

5 PHENOLIC AND ENOLIC ANTIOXIDANTS

I. REDOX REACTIONS OF PHENOLS

A. One-Electron Oxidations of Phenols

As intrinsically electron-rich compounds, phenols are prone to enter into efficient electron-donation reactions with oxidizing agents. Typically, an electron is transferred from a phenol to the unfilled orbital of a one-electron oxidant such as a peroxyl radical, followed by rapid proton transfer (Equation 5-1):

$$Ph\text{-}OH + ROO\cdot \rightarrow Ph\text{-}O\cdot + ROOH$$

(Equation 5-1)

The net result is the equivalent of a hydrogen atom transfer from the phenolic hydroxyl group to the free radical. The product, the phenoxyl radical (or phenolate radical), PhO·, is stabilized by resonance delocalization of the unpaired electron to the *ortho* and *para* positions of the ring (Equation 5-2). Therefore, the phenoxyl radical possesses radical density at both oxygen and carbon centers.

In order to act as an effective antioxidant, the phenoxyl radical must not be reactive enough to initiate further free-radical reactions on its own. Whereas phenol itself is a rather ineffective antioxidant, when other radical-stabilizing features are accentuated, extremely potent antioxidants can result. In addition to the intrinsic stability of the radical that results from resonance properties, these characteristics can be intensified by various structural features, such as steric and inductive effects of ring substituents. For example, bulky or electron-donating groups at the *ortho* positions of a phenol characterize a number of efficient antioxidants, both synthetic and naturally occurring. Phenoxyl radicals are, however, reactive enough to oxidize ascorbic acid and to reduce some quinones (Neta and Steenken, 1981).

Phenoxyl radicals can also be stabilized by hydrogen bonding; for example, an adjacent hydroxyl or amino group, as found in catechols or catecholamines, will augment the stability of a derived radical (5-1).

(5-1)

A further feature of the structure of phenols is their acidity, or ability to ionize; the phenolate anion is even more readily oxidized than the protonated form. (Phenols, because the anion is resonance-stabilized, are many orders of magnitude stronger as acids than aliphatic alcohols.) The pKa's of phenolic compounds vary widely; phenol's is 9.9, but others are much stronger acids. Introduction of electron-withdrawing groups such as nitro or chloro greatly lowers the pKa of the resulting phenol. That for *p*-nitrophenol, for example, is 7.2, and 2,4,6-trinitrophenol has such a low pKa (approximately 0) that it has the trivial name of "picric acid." However, even many polyphenols (where the ring substituents would be considered electron-donating) display lowered pKa's; numerous flavonoids, for example, have first pKa's in the pH 6.0–7.0 range (Jovanovic et al., 1994). Presumably, this is due to the stabilization of the anion by an adjacent –OH substituent. Therefore, it would be expected that these compounds would be readily oxidized at pH's near neutrality, and this assumption is largely borne out.

The various structural features of different phenols impart very different reactivities toward oxidants of moderate activity. For example, their rate constants for reaction with trichloromethylperoxyl radicals vary by over three orders of magnitude (Table 5-1).

B. Reactions of Phenoxyl Radicals

One-electron reactions of certain phenoxyl radicals include disproportionation, in which a pair of radicals undergo self-redox conversion to even-electron products (Reaction 5-2). This reaction is especially characteristic of hydroquinones and other polyphenols. The intermediate phenoxyl radical, or semiquinone, after acceptance of another electron, is at the oxidation level of a phenol, and after protonation becomes one. After the transfer, the donor radical is at the quinone oxidation state, requiring only orbital reorganization to achieve this formal structure.

Table 5.1 Second-Order Rate Constants for Substituted Phenols with the $Cl_3COO\cdot$ Radical

Phenol	Rate constant
Phenol	$<1 \times 10^5$
Gallic acid	4.5×10^5
Thymol	3.8×10^6
Carvacrol	3.9×10^6
Ferulic acid	5.6×10^6
(+)-Catechin	7.5×10^6
Quercetin	3.9×10^7
Trolox	1.7×10^8
Morin	3.0×10^8
Fisetin	4.1×10^8

Note: Units: $L\ mol^{-1}\ sec^{-1}$.

From Scott et al. 1993; Aruoma, 1993; Aeschbach et al., 1994.

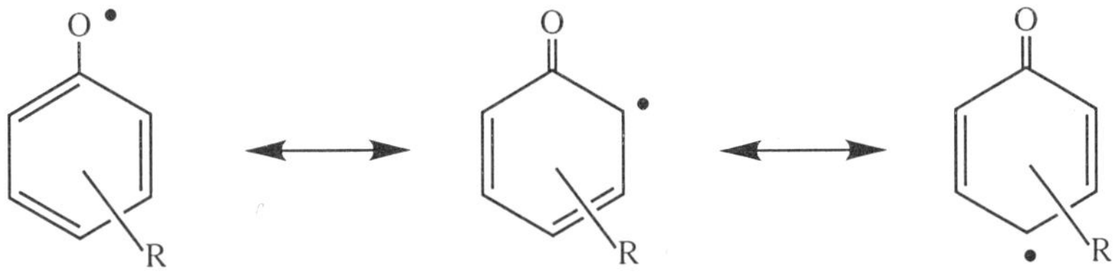

(Equation 5-2)

Phenoxyl radicals also undergo dimerization ("phenol coupling," Equation 5-3) to produce new C–C or C–O bonded hydroxylated biphenyls or hydroxydiphenyl ethers. These reactions have been observed in a wide variety of environmental and biological media (Taylor and Battersby, 1967). They are probably of biosynthetic importance, particularly in the case of complex alkaloids.

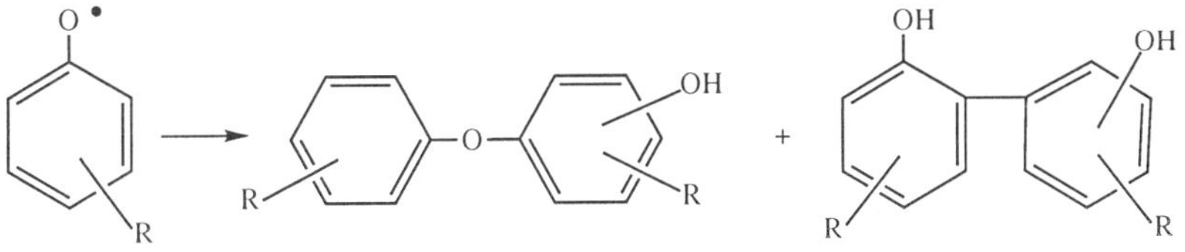

(Equation 5-3)

The reactions of phenoxyl radicals with external odd-electron species are often of paramount importance for their antioxidant activity. For example, a phenoxyl can add a peroxyl radical to give an even-electron product, a peroxycyclohexadienone (Equation 5-4):

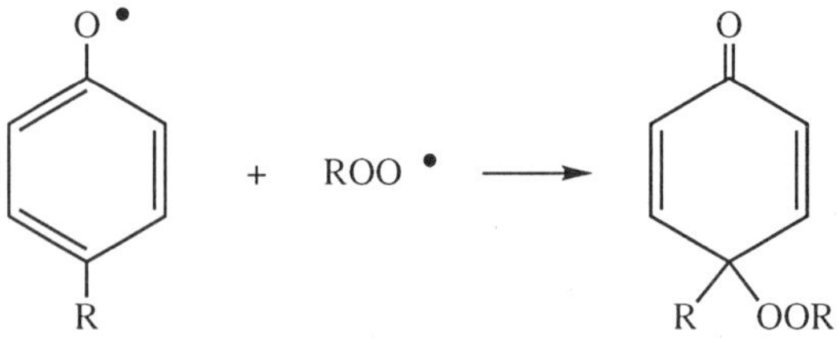

$$\text{[phenoxyl radical]} + ROO^\bullet \longrightarrow \text{[cyclohexadienone-OOR adduct]}$$

(Equation 5-4)

Since the original phenoxyl radical may also have been formed by attack of a peroxyl radical, the reaction with a second molecule represents a particularly efficient process for deactivating a radical chain reaction. The stoichiometry of two ROO· per phenol has been observed in many inhibited chain processes.

C. Phenols as Electron Donors

Phenols, as a compound class, have repeatedly been shown to donate electrons more or less efficiently, and for almost half a century the strong inverse relationship between the oxidation potential of a phenol and its antioxidant efficiency has been recognized (Bolland and Ten Have, 1947). Because of their importance as synthetic antioxidants, the redox properties of synthetic phenolic compounds have been investigated repeatedly by various techniques. Table 3-1 contains some of these values. As expected, the more easily oxidized phenols tend to be those with electron-donating groups attached to the ring. For example, catechol (*o*-dihydroxybenzene) has a half-wave potential of +530 mv (Scurlock et al., 1990); phenol itself is around +900; and *p*-nitrophenol, +1170 (Stone, 1987).

The redox equilibria and reactions of naturally occurring phenols have not been as thoroughly investigated. Using a pulse radiolysis technique, the one-electron redox potentials of various phenols and amines of biological interest (including many which have been reported to show antioxidant activity) were determined (Steenken and Neta, 1982). The redox potentials of many flavonoids (polyphenols such as quercetin, 5-2), and chalcones have been determined by pulse radiolysis (Jovanovic et al., 1994), cyclic voltammetry (Hodnick et al., 1988) and discontinuous titration (Loth and Diedrich, 1967). The half-wave

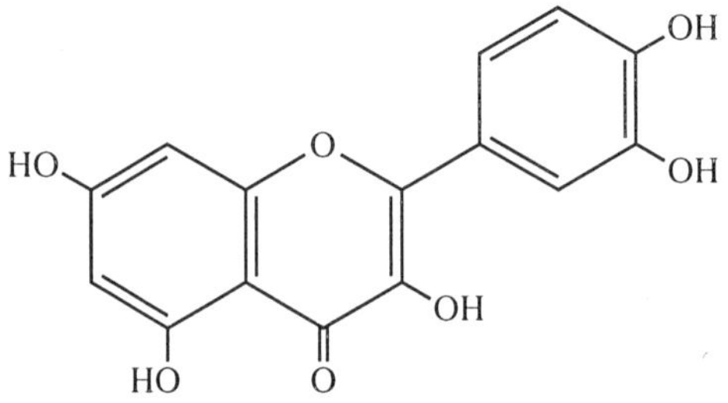

(5-2)

Table 5.2 Second-Order Rate Constants for Substituted Phenols with the HO· Radical

Phenol	Rate constant
Phenol	9×10^9
Catechol	1.1×10^{10}
o-Cresol	1.1×10^{10}
p-Coumaric acid	8.2×10^9
p-Hydroxybenzoic acid	8.5×10^9
Sinapic acid	2.2×10^{10}
Syringic acid	1.6×10^{10}
p-Nitrophenol	3.8×10^9
Tyrosine	1.3×10^{10}
Procyanidin B2	1.4×10^9
Eugenol	$10^9 - 10^{10}$
Vanillic acid	1.6×10^{10}
Ferulic acid	4.6×10^9
(+)-Catechin	3×10^9

Note: Units: L mol⁻¹ sec⁻¹.

From Buxton et al., 1988; Ricardo da Silva et al., 1991; Taira et al., 1992; Scott et al., 1993.

potentials of 17 vitamin E derivatives, and of 11 hydroquinones, have also been tabulated (Mukai et al., 1991, 1993c).

II. REACTIONS OF PHENOLS WITH OXIDIZING FREE RADICALS

A. Reactions with Hydroxyl Radical

Rate constant data for reactions of naturally occurring phenolic compounds with free radical oxidants are sparse. The relative (if not the absolute) efficiencies of quenching of ·OH radical have been determined for several classes of naturally occurring compounds with possible antioxidant activity, including polyphenols (Husain et al., 1987; Chimi et al., 1991; Ricardo da Silva et al., 1991) and phenylpropanoids (Taira et al., 1992; Scott et al., 1993). All of the rate constants are, as expected, quite large. The data are summarized in Table 5-2.

B. Reactions with Superoxide

Superoxide reactions with phenols appear to be inversely correlated with their one-electron redox potentials. This suggests that it is oxidizing the compounds, unlike other common reactions with supeoxide that are reductions. Phenol itself, for example, with its rather high redox potential of ca. 900 mV (Table 3-1), was oxidized at a rate of 580 l/mol sec (Tsujimoto et al., 1993). A polyphenol, rutin, whose redox potential is only 600 mV, had a rate constant that was about an order of magnitude faster (Jovanovic et al., 1994). The assumption, supported by considerable evidence, is that the first intermediate in the reaction is the phenoxyl radical.

The superoxide radical, $\cdot O_2^-$, also reacts further (under at least some conditions) with phenoxyl radicals by addition. Jin et al. (1993) studied the reaction of the tyrosyl radical with superoxide. Previous to their investigation, it was thought that electron transfer from superoxide to tyrosyl was predominant, since the redox potentials of the two species made such a reaction favorable. Product and kinetic studies, however, provided good evidence for the following pathway (Equation 5-5):

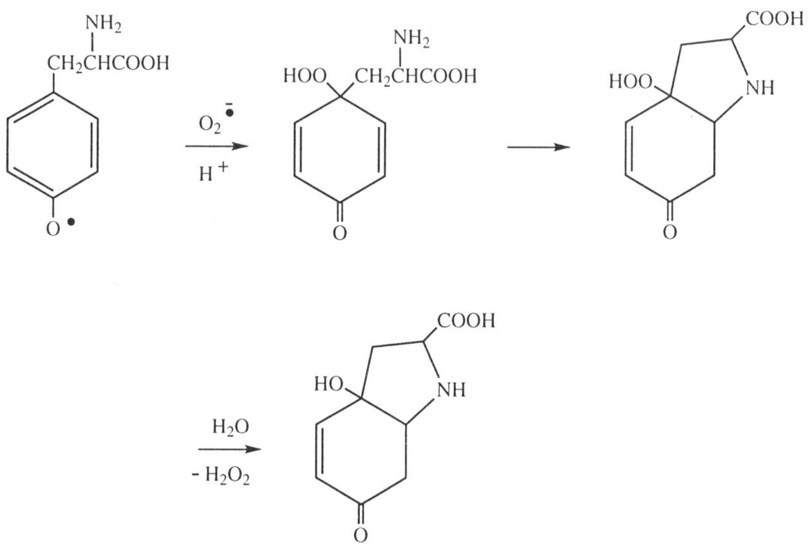

(Equation 5-5)

A number of phenols and phenolic antioxidants such as Vitamin E interact with superoxide at widely varying rates (Tsujimoto et al., 1993) to produce an intermediate shown by ESR studies to be the phenoxyl radical (Tajima et al., 1983). (Others, however, believe that the first intermediate in the reaction is an addition product: Cadenas, 1995.) There is little evidence that simple phenoxyl radicals undergo further attack by superoxide (although polyphenols may). In the case of a hindered antioxidant such as BHT, all the *o* and *p* positions of the ring are substituted, unlike the tyrosine situation described above, which would probably tend to slow addition reactions. The rate constant for the reaction of vitamin E with HOO·, the protonated form of superoxide, was reported to be 2×10^5 l/mol sec, about two orders of magnitude faster than the corresponding HOO·-unsaturated lipid reaction (Bielski, 1983).

Flavonoids have also been shown repeatedly to react with $\cdot O_2^-$ generated by a variety of methods (Robak and Gryglewski, 1988; Ricardo da Silva et al., 1991; Cotelle et al., 1992; Scott et al., 1993; Jovanovic et al., 1994). The latter authors showed that they reacted at rather moderate rates, with second-order rate constants ranging from about 300 l/mol sec (for 4′,5,7-trihydroxyflavone, [5-3]) to 5×10^4 l/mol sec (for rutin, 5-4). The fastest reactions occurred with those flavonoids having the most extensive B-ring hydroxylation. The reactions were oxidations of the flavonoids, as confirmed by the isolation of H_2O_2 (the reduction product of superoxide) from the reaction mixtures in nearly quantitative yield. This would be consistent with the formation of the phenoxyl radical as an initial step in the reaction. The weight of the evidence seems, then, to support the role of superoxide as an oxidizing agent in these systems, despite the unfavorable thermodynamics of direct electron transfer. These observations, and those of the vitamin E reactions mentioned earlier, may represent an example of sequential proton-electron transfer (Sawyer et al., 1985) or one-electron transfer occurring concurrently with proton transfer in the transition state (Neta et al., 1989).

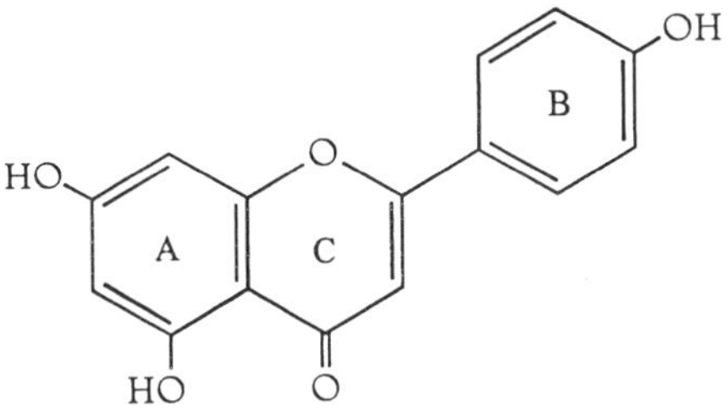

(5-3)

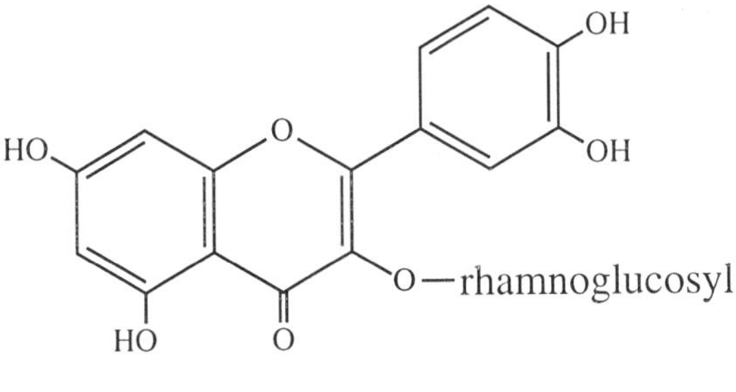

(5-4)

For these compounds, their scavenging activity toward superoxide did not correlate at all with their ability to inhibit lipid peroxidation in a mouse liver homogenate (Yuting et al., 1990). This suggests that other mechanisms of inhibition could be more important for polyphenols, at least in this system. Kaempferol (3,4′,5,7-tetrahydroxyflavone) underwent photoreaction with riboflavin, via a pathway that may have involved superoxide as an intermediate,

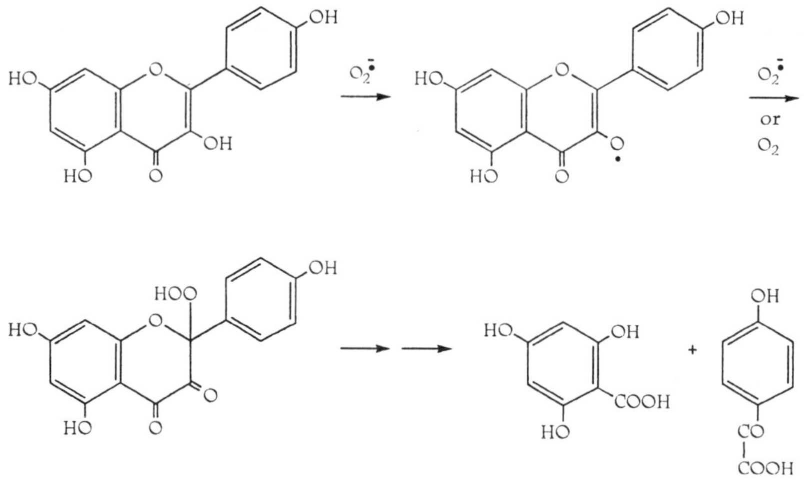

Figure 5-1 Proposed mechanism of riboflavin-induced cleavage of kaempferol.

to cleavage products that included hydroxybenzoic acids (Figure 5-1) (Takahama, 1987). Again, it is likely that the first step of the reaction was the formation of the phenoxyl radical.

Many flavonoid phenoxyl radicals appear to undergo subsequent one-electron oxidations, possibly with molecular oxygen or peroxyl radicals, to afford quinones (usually *o*-quinones). These products may be oxidized further to ring-opened products (Uri, 1961; Nordström, 1968; Takahama, 1987).

C. Reactions with Peroxyl Radicals

The kinetics of reactions between synthetic phenols and peroxyl radicals have been investigated for decades due to their importance as antioxidants for foodstuffs, rubber, and other materials (see Chapter 4). Simple phenolic compounds show the expected trends in reactivity for an oxidation reaction, with the more electron-rich compounds generally reacting more readily; reactivity correlates with such measures of electron density as the Hammett σ constant, for example.

Although naturally occurring polyphenols have repeatedly been shown to scavenge peroxyl radicals (Torel et al., 1986), few rate constants for inhibition have been measured, except for vitamin E, which usually gives a value of around 10^6 l/mol sec (Niki and Matsuo, 1992). A few measurements for the highly reactive trichloromethylperoxyl radical are summarized in Table 5-1; the numerical values for these rate constants are, as expected, considerably higher. The trend of reactivity, however, should be similar for other peroxyl radicals.

More recently, data for the reactions of peroxyl radicals with flavonoids have begun to appear. Using the induction-period method, Ariga and Hamano

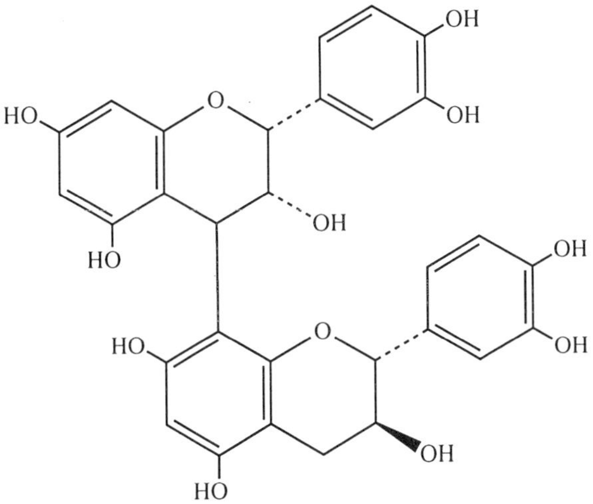

(5-5)

(1990) measured k_{inh} in aqueous solution of 6×10^4, 5.9×10^4, and 2.7×10^4 l/mol sec for procyanidin B-1 (5-5), procyanidin B-3 (5-6), and (+)-catechin (5-7), respectively. For comparison, k_{inh} for vitamin E was 1.1×10^5 l/mol sec in their system. Since polyphenols may have numerous potential sites for reaction with peroxyl radicals, they may have high stoichiometries, far above the typical 2.0 radicals per antioxidant molecule for a synthetic antioxidant such as BHT: values of 8 or more were measured for the procyanidins 5-5 and 5-6 (Ariga and Hamano, 1990). Belyakov et al. (1995) have measured the

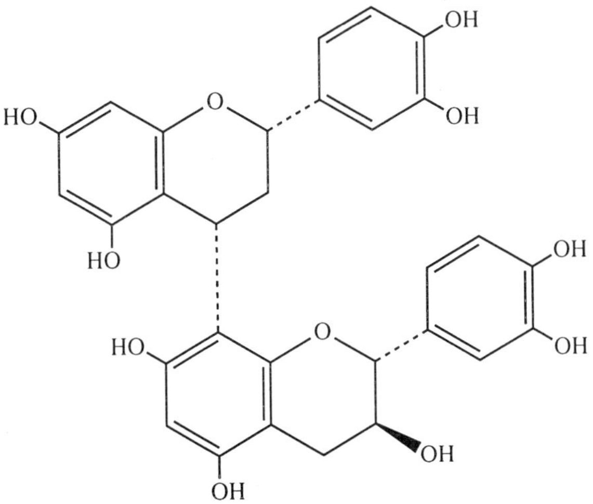

(5-6)

(5-7)

rate constants for the reactions of various flavonoids and related compounds with the peroxyl radical $Ph_2CHOO\cdot$, using chlorobenzene as solvent and a reaction temperature of 50°C. They monitored the rate constants by the use of chemiluminescence. Surprisingly high values were determined for many of the compounds (Table 5-3), that were comparable to or even greater than that for vitamin E, which was equal to 8.5×10^6 l/mol sec in this system. The authors suggested that the primary phenoxyl radicals observed in their studies might be rapidly transformed into substances with lesser activity or even pro-oxidant character, reducing their usefulness in practical systems.

III. REACTIONS OF PHENOLS WITH SINGLET OXYGEN

Many phenolic compounds, in addition to being potent quenchers of free radical reactions, also react quite rapidly with $^1O_2$. Vitamin E, for example, has been determined by several authors (Wilkinson and Brummer, 1981) to

Table 5-3 Second-Order Rate Constants (in Chlorobenzene at 50°C) of Naturally Occurring Phenolic Compounds with the $PhCH_2OO\cdot$ Radical

Compound	Rate constant
Naringenin	3.4×10^3
Kaempferol	1×10^6
Catechin	6.6×10^6
Vitamin E	8.5×10^6
Nordihydroguiaretic acid	1.0×10^7
Fisetin	1.2×10^7
Caffeic acid	1.5×10^7
Dihydroquercetin	1.9×10^7
Quercetin	2.1×10^7
Luteolin	2.2×10^7

Note: Units: L mol^{-1} sec^{-1}.
From Belyakov, V. A., et al., 1995, *J. Chem. Soc. Perkin Trans.*, 2, 419–444.

have a rate constant (in alcoholic solvents) of roughly 5×10^8 l/mol sec (about 5% of that of β-carotene) with $^1O_2$. This is a numerically greater rate constant than all but a few naturally occurring compounds possess, and it is several orders of magnitude greater than that for most biomolecules that make up such important polymers as DNA, proteins, and lipids. Interestingly, there is a high degree of correlation between the $^1O_2$ quenching activity and radical-quenching activities of 17 vitamin-E related compounds (Mukai et al., 1991).

Some flavonoids, too, which are also well-known radical scavengers, appear to be reactive with $^1O_2$ (Sorata et al., 1984; Wagner et al., 1988; Tournaire et al., 1993). Most of these data have been conducted under conditions in which $^1O_2$ is not the sole reactant, or in model media such as liposomes, and accordingly only rather qualitative statements concerning their reactivities have been feasible. Recently, however, the kinetics of $^1O_2$ quenching by a number of flavonoids of various structural types have been systematically studied. The most chemically reactive quencher types were flavonols such as quercetin (5-2) and fisetin (5-8), with rate constants for reaction with $^1O_2$ of $2-3 \times 10^6$ l/mol sec, about 2 orders of magnitude lower than vitamin E (Tournaire et al., 1993; Briviba and Sies, 1994). These rate constants, however, are greater than those for many biomolecules such as unsaturated fatty acids which are often thought of as being susceptible to $^1O_2$ damage, so it is possible that certain flavonoids might provide protection if they were present in sufficient concentration.

(5-8)

The products of these reactions have not often been fully characterized, but in some cases (for example, that of catechin, whose rate constant for quenching of $^1O_2$ is about 6×10^6 l/mol sec) physical quenching again seems to predominate (the flavonoid quenches singlet oxygen without undergoing much change in concentration). In other cases, however, such as the flavonols mentioned above, the flavonoid is removed rapidly from solution (Tournaire et al., 1993). Of the several type reactions of flavonols and $^1O_2$ that have been reported, the most favorable appears to be an opening of ring C (Matsuura and Matsushima, 1970), probably via an unstable oxetane derivative (Figure 5-2), leading to the formation of an ester.

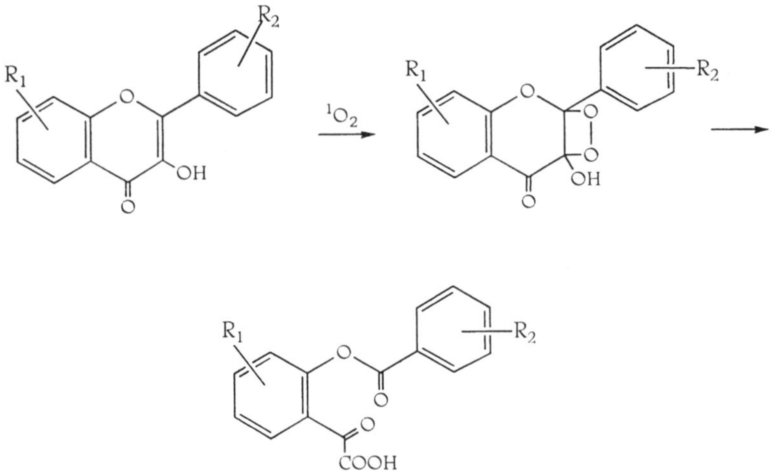

Figure 5-2 Proposed mechanism of singlet oxygen-induced cleavage of flavonols.

The presence of an OH group on the C ring double bond undoubtedly activates it toward attack by the electrophile, $^1O_2$. If the OH is replaced by an O-sugar group (as in the glycoside rutin 5-4, for example), reactivity is decreased by about a factor of 2–3 (Tournaire et al., 1993; Briviba and Sies, 1994).

Mukai et al. (1993c) report that hydroquinones such as ubiquinols scavenge singlet oxygen at surprisingly high rates. The rate constant for ubiquinol-10 (cf. 5-15), the natural coenzyme Q reduction product, was the highest (1.6 × 10^8 l/mol sec) of all hydroquinones tested (it was more than ten times as reactive as hydroquinone itself). Other authors, however, have reported a rate constant about one-fifth as great (Briviba and Sies, 1994). For all the quinones tested, the rate constants correlated reasonably well ($r = -0.87$) with the redox potential as measured by cyclic voltammetry. The rate constant for α-tocopherol, as measured by these authors, was only about 25% greater (Mukai et al., 1993c).

IV. CLASSES OF PHENOLIC AND ENOLIC ANTIOXIDANTS

A. Vitamin E and Related Compounds

Tocopherols and related substances (tocols, tocotrienols) are compounds found in high concentrations (up to 1000 ppm) in certain vegetable oils, grains, and other plant products. Wheat germ and maize, sunflower, and cottonseed oils are especially rich in these compounds, which are extremely water-insoluble. In the intact plant, they are found in chloroplasts (where there may be as many as one molecule of α-tocopherol for every ten molecules of chlorophyll) as well as in other lipid-rich environments. They are also present, at much lower levels, in animal tissues; the human heart, for example, contains about 20 ppm of α-tocopherol. In the animal body they are also usually localized in sites rich in lipids, such as blood lipoproteins, microsomes, or mitochondrial membranes.

Discovery of the compound class came more than 75 years ago. The compounds have since been reported to display diverse and pronounced physiological activities, including roles as antisterility agents, stimulators of nerve transmission, inhibitors of platelet aggregation, and, perhaps most importantly, potent antioxidants for cell membranes and other important biological assemblages. They are also used as antioxidants for foodstuffs such as fats and oils, either by themselves or, more typically, as a mixture together with ascorbate, citrate, gallate, or lecithin derivatives.

Many isomers of tocopherols have been described. The α-, β-, γ-, and δ-isomers (5-9–5-12) are usually present in the greatest abundance. The parent compound, α-tocopherol or vitamin E, features three methyl groups and a *p*-ether linkage in the phenolic ring. Related isomers differ slightly as indicated. The ester α-tocopheryl acetate also occurs naturally in some plants; it has virtually no antioxidant activity because of the lack of the phenolic hydroxyl group, but evidence suggests that the ester can be hydrolyzed under some conditions. The concentration of vitamin E in mammalian cell membranes, although quite low in absolute terms (<0.1 nmol/mg protein) (Packer and Landvik, 1989), nevertheless appears to be adequate to prevent most instances of autooxidative damage in normally functioning cells. The order of antioxidant activity among the tocopherols is ordinarily α- > β- > γ- > δ.

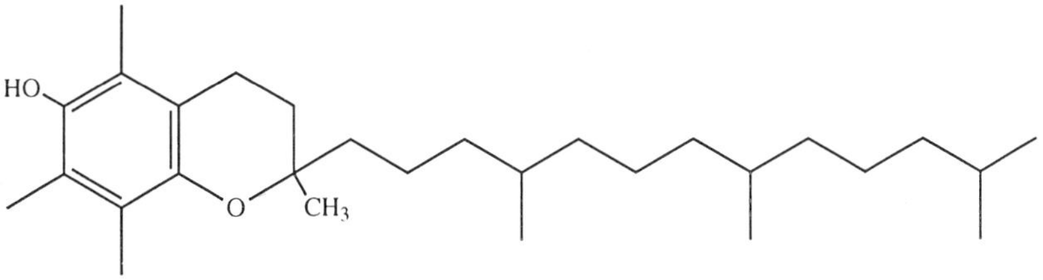

(5-9)

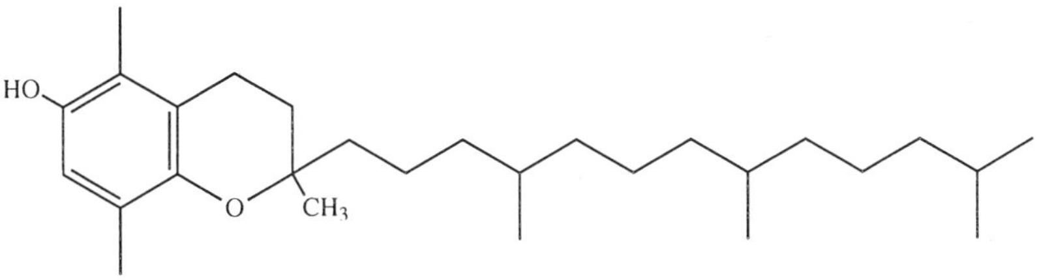

(5-10)

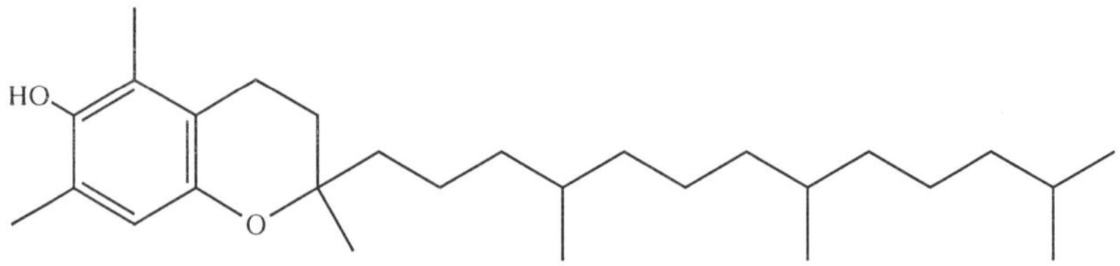

(5-11)

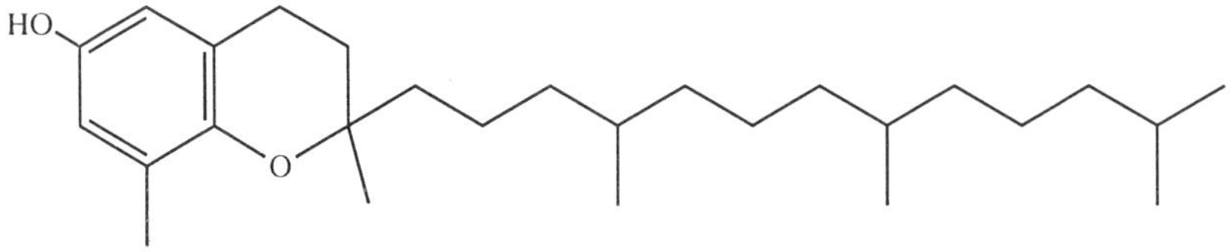

(5-12)

Although vitamin E is an excellent quencher of singlet oxygen, and reacts with HO· at about the same rapid rate as many other biomolecules, its antioxidant activity *in vivo* is almost certainly due to its scavenging of peroxyl and alkoxyl radicals. It has been repeatedly shown to react very rapidly with peroxyl radicals; its rate constants with these species are usually agreed to be the highest of all naturally occurring substances (Burton et al., 1983). This is the basis for its outstanding ability to act as a chain-breaking antioxidant (Chapter 2).

The initial reaction of a peroxyl radical with vitamin E produces the phenoxyl radical (by either electron- or hydride-transfer reactions), and a hydroperoxide by the mechanism of reaction 5-1, and a subsequent reaction with another peroxyl radical gives rise to addition products such as 5-13. These processes largely account for the observed stoichiometry of approximately 2 ROO· quenched per molecule of vitamin E which has been observed by many investigators of this reaction. In addition, mechanisms are known for "repairing" the vitamin E free radical, that is, for donating an electron to it to restore it to the even-electron form. Vitamin C is the best-known of these electron

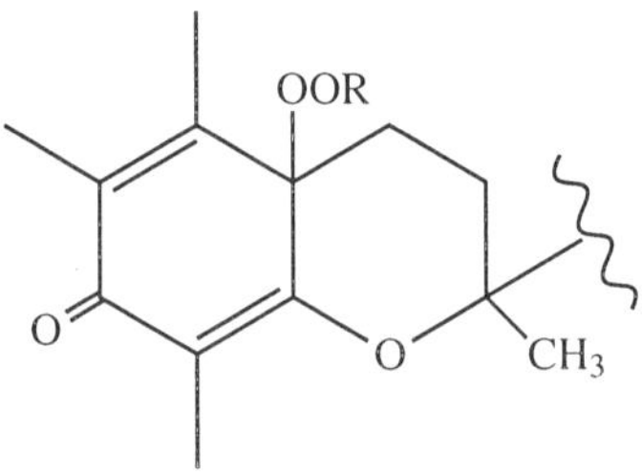

(5-13)

donors. Such mechanisms would have the potential for making vitamin E a catalytically active antioxidant, allowing one molecule to deactivate many free radicals. (Another species, such as vitamin C, would have to be stoichiometrically consumed, of course.)

Many other oxidation products of vitamin E have been observed in *in vitro* and *in vivo* systems. Some of these include quinones (such as α-tocopherylquinone, 5-14), dimers, and quinone-epoxides (Schuler, 1990). It appears that some animal tissues contain an enzyme that reduces the tocopherylquinone to the corresponding hydroquinone, which is still an effective antioxidant (Bindoli et al., 1985; Hayǎshi et al., 1992).

(5-14)

Although reaction products of Vitamin E with $^1O_2$ have been determined, the bulk of the quenching appears to be physical (Fahrenholtz et al., 1974), although some authors have proposed a second pathway in which it is irreversibly oxidized to quinones and quinone epoxides (Fryer, 1992).

For many years some anomalous properties of vitamin E, namely its ability to act as a promoter of peroxidation reactions under some conditions, have remained unexplained. Recently, however, Ingold and co-workers (1993) suggested an ingenious hypothesis to account for these results. They propose that the vitamin E phenoxyl radical may be able to promote lipid oxidation in some environments, such as low-density lipoprotein dispersions, in which restricted diffusion prevents the restoration of the starting phenol by hydrogen transfer from an efficient donor such as vitamin C or ubiquinol (5-15). Mukai et al. (1993b) provided additional evidence for this pathway by measuring the rate constants for hydrogen abstraction from various lipid substrates by the tocopheroxyl radical. The values, while low (<0.1 l/mol sec) could nevertheless allow for some vitamin E-promoted lipid peroxidation in the absence of competing reactions. In a more recent study, Foti et al. (1996) have claimed that phenoxyl radicals have very high rate constants for H-atom abstraction from phenols (up to 300× greater than those for peroxyl radicals) when the large Arrhenius preexponential factors for these reactions are taken into account. This phenomenon might have important consequences for the ability of vitamin E and similar phenolic antioxidants to protect substances with labile hydrogens, such as proteins, from oxidative damage.

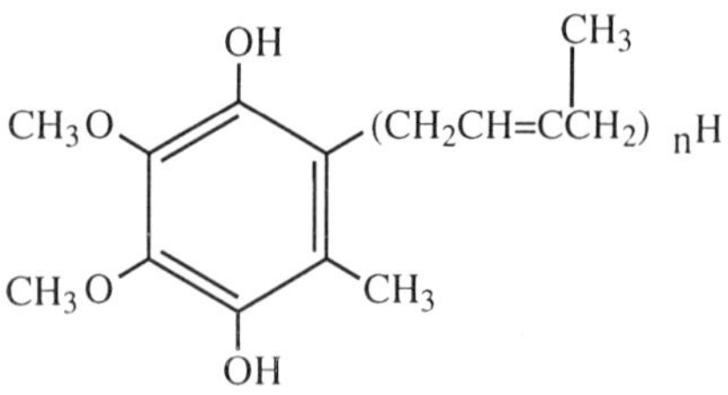

(5-15)

Vitamin E analogues are also good antioxidants. Osawa et al. (1991) identified interesting tetracyclic compounds from the leaf wax of *Prunus grayana,* prunusols A and B (5-16–5-17), which contain structural elements derived from a tocopherol and a cinnamic acid. Although neither of the moieties would be anticipated on structural grounds to exhibit particularly strong antioxidant activity, in tests with autooxidizing linoleic acid in ethanol, the compounds showed inhibition (at about 6×10^{-4} M) that was comparable to BHA and other tocopherol derivatives.

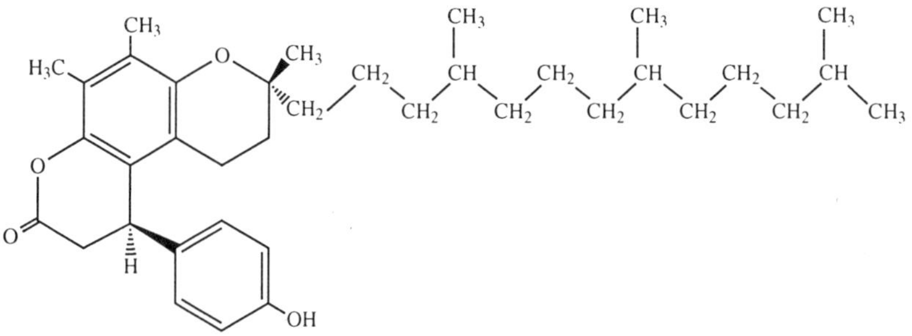

(5-16)

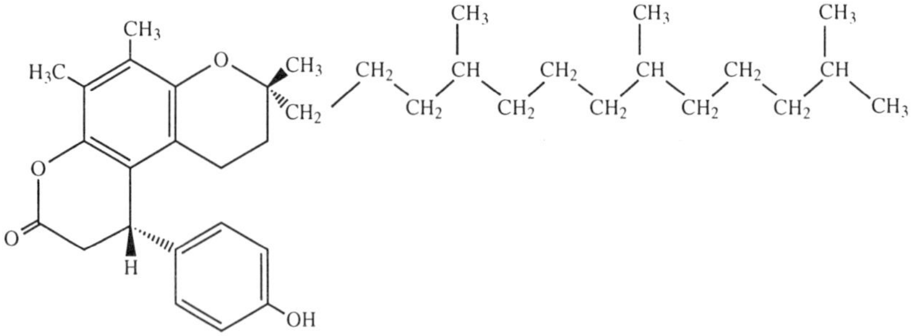

(5-17)

B. Vitamin C

Vitamin C (ascorbic acid, AH: 5-18) is found in quite high (millimolar and up) concentrations in the aqueous fractions of many animal tissues including the spinal cord, lung, and eye. In plants, some fruits may contain more than 1% (~6 mM). In human blood plasma it normally occurs at a level of about 0.1 mM. Although most organisms are able to synthesize it, a few (including humans), must obtain it in their diets. Because of its enediol structure, it displays a rather low first pKa (about 4.2) and accordingly exists almost entirely as the monoanion in most tissues. The hydrogen at the 3-position, which is the most acidic one, is also the hydrogen atom removed in one-electron oxidation reactions.

(5-18)

Ascorbic acid has been used as an antioxidant in a great variety of products, and is extensively used in the food industry. It synthetic isomer, erythorbic acid, which differs only in the configuration of one side-chain hydroxyl group, is also a powerful antioxidant but, paradoxically, has little vitamin activity. It appears, however, also to occur naturally and to take the place of ascorbate in certain fungi (Nick et al., 1986).

Ascorbic acid is a potent reducing agent; its low redox potential of about 280 mV (Table 3-1) means that it has the thermodynamic potential to react with almost all other oxidizing free radicals. Experimentally, it reacts rapidly with even weakly oxidizing agents, including the free radicals $\cdot O_2^-$ and $\cdot OOH$, as shown by pulse radiolysis and flash photolysis studies (Cabelli and Bielski, 1983). The rate constant for reaction of superoxide with ascorbate at near-neutral pH has been reported to be in the range of $2.7 \times 10^5 - 6.5 \times 10^5$ l/mol sec, which is greater than that for any other biomolecules except glutathione, metalloproteins, a few polyphenols, and quinones (Halliwell and Gutteridge, 1989; Niki, 1991). The product, ascorbate radical ($\cdot A^-$, also known by the cumbersome and inaccurate appellation "semidehydroascorbic acid"), can terminate a chain reaction by disproportionating into the nonradical products, ascorbate and dehydroascorbate. However, the reaction efficiency is partly ameliorated by the ability of ascorbate to produce superoxide upon its own oxidation by molecular oxygen:

$$AH + \cdot O_2^- \rightarrow \cdot A^- + O_2 + H^+$$

(Equation 5-6)

$$AH + O_2 \rightarrow \cdot A + \cdot O_2^- + H^+$$

(Equation 5-7)

The presence of trace amounts of transition metal ions as catalysts for these reactions appears to govern the relative extents of the two processes (Bendich et al., 1986, Buettner, 1988). Iron and copper salts are well known to promote the formation of H_2O_2 and hydroxyl radicals from ascorbate, which probably accounts for the occasional reports that it can be cytotoxic or act as a pro-oxidant under some conditions (Bendich et al., 1986; Niki, 1991).

Dehydroascorbate can also be formed directly in a two-electron, non-enzymatic oxidation promoted by hydrogen peroxide (Groden and Beck, 1979). This reaction appears to be highly significant *in vivo*.

Hydroxyl radical reacts with ascorbate, as it does with other olefinic substrates, by addition as well as by H-abstraction. Several free-radical products that originate from such addition to the double bond have been characterized by ESR spectroscopy (Fessenden and Verma, 1978). The further chemistry of these radicals is quite variable, but in any event their formation serves an antioxidant function, removing a highly potent oxidant and converting it to less reactive forms.

Ascorbic acid reacts with typical peroxyl radicals in water with a rate constant of around 2×10^6 l/mol sec (Simic, 1988), and with the more active trichloromethylperoxyl radical nearly 100 times faster (Aruoma, 1993). (Others, however, reported an inhibitor rate constant of 5×10^4 l/mol sec; Ariga and Hamano, 1990). This value is approximately a factor of 4 lower than the vitamin E rate constant. The number of peroxyl radicals trapped per mole of ascorbate ranged from near 2 at low ascorbate concentrations to almost 0 at high concentrations (Bendich et al., 1986), apparently because of self-termination reactions that compete with radical trapping at higher ascorbate levels.

Ascorbic acid quenches singlet oxygen with a rate constant of about 10^7 l/mol sec (Chou and Kahn, 1983; Rougée and Bensasson, 1986). The products include oxalic acid, $(HOOC)_2$, and threonic acid, $HOOC-(CHOH)_2-CH_2OH$ (Schenck, 1960). It is probable that the quenching ability of ascorbate is not particularly important in the aqueous environment where it is found, since $^1O_2$ has such a short lifetime in water in any case.

C. Flavonoids and Derivatives

1. Flavonoids

Flavonoids represent a large and diverse group of phenolic compounds derived from higher plants. Derived from the type structure, flavone (5-19), these heterocyclic compounds display a wide range of substitution patterns and oxidation states including flavonols (5-20), flavanols (5-21), flavanones (5-22), and flavans or catechins (5-23). The hydroxylation and alkoxylation

PHENOLIC AND ENOLIC ANTIOXIDANTS

patterns of the A and B rings of these compounds vary extensively and are of great importance in determining their activity as antioxidants. In addition, the substitution of an OH group at the 3-position of the C ring is also of importance for these properties.

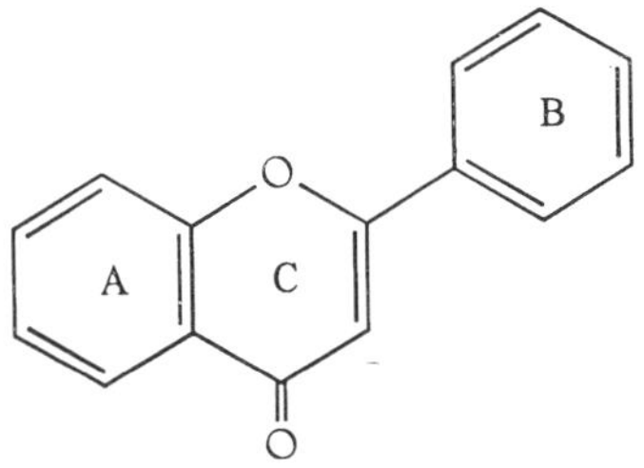

(5-19)

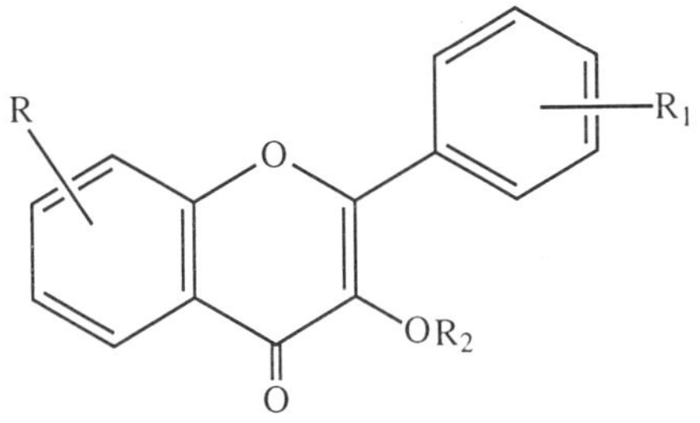

(5-20)

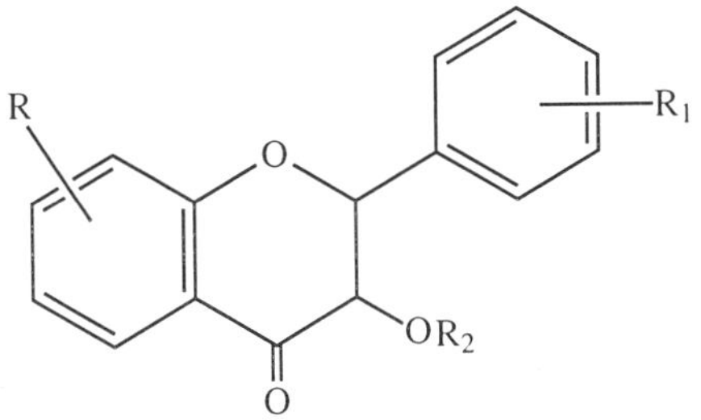

(5-21)

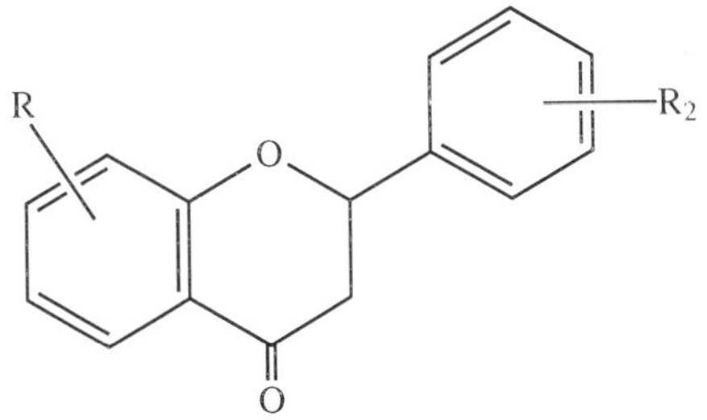

(5-22)

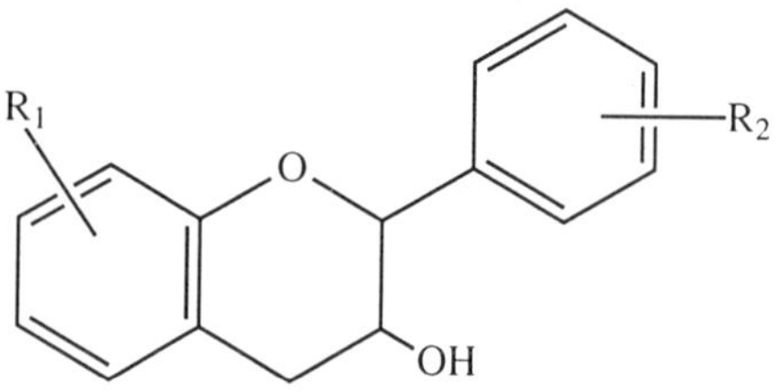

(5-23)

The ease of oxidation and the antioxidant activity of many flavonoids and derivatives has been demonstrated in a variety of oxidizing systems for many years. The compounds appear to possess a variety of mechanisms of action which include radical scavenging and metal ion complexation. The ability of quercetin (5-2) to inhibit the copper-catalyzed degradation of lard, for example, has been attributed to be at least in part due to its metal complexing ability (Mahgoub and Hudson, 1985). Quercetin formed the most stable complexes with Cu(II) of all flavonoids tested (Thompson et al., 1976; Takamura and Ito, 1977). Structural features required for complex stability included the 3-hydroxy-4-keto grouping in ring C (flavonol structure).

Repeated studies have shown that flavonoids having greater numbers of hydroxyl groups, or hydroxyl groups localized *ortho* to one another, are more effective antioxidants. The B ring of most flavonoids is usually the initial target of oxidants, as it is more electron-rich than the A and C rings, whose electron densities are somewhat drained away by the carbonyl group. These properties are consistent with the expected mechanisms of oxidation of phenols; electron-donating substituents, such as hydroxyl groups, should lower the oxidation potential for a compound, and *ortho* hydroxylation should stabilize phenoxyl radicals, as described earlier (Reaction 5-1). As might be expected, glycosylation of one or more of the hydroxyl groups of a flavonoid greatly reduces its antioxidant activity.

Flavonoid compounds having both *o*-hydroxylation in the B ring and multiple hydroxylation in the A ring, such as quercetin (5-2), robinetin (5-24),

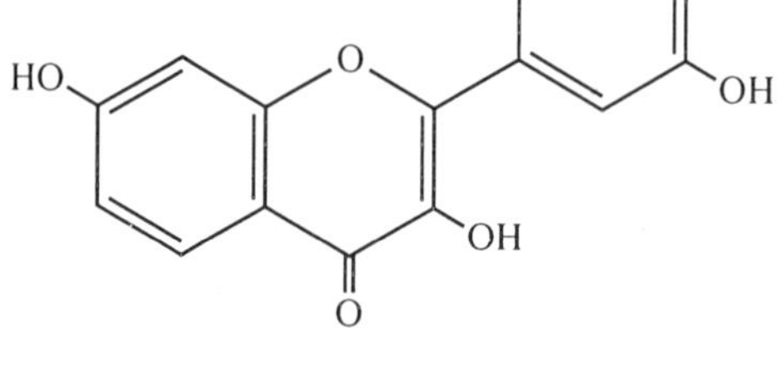

(5-24)

luteolin (5-25) and myricetin (5-26), have been demonstrated to be particularly effective antioxidants in many studies (Dziedzic and Hudson, 1983b; Pratt and Hudson, 1990; Rice-Evans et al., 1995). Compounds lacking these features, such as kaempferol (5-27), naringenin (5-28), hesperetin (5-29), and apigenin (5-30) are usually not nearly as efficient (Pratt and Hudson, 1990). Isovitexin (5-31), a C-glycosyl flavonoid isolated from rice hulls, may be an exception to this rule, but the reasons for its reported high activity are not clear (Ramarathnam et al., 1989).

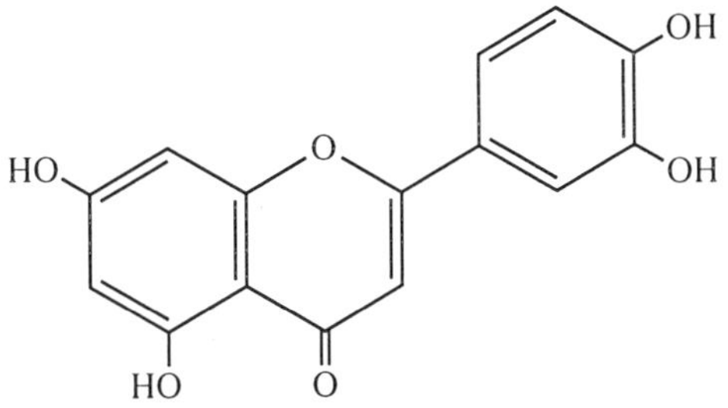

(5-25)

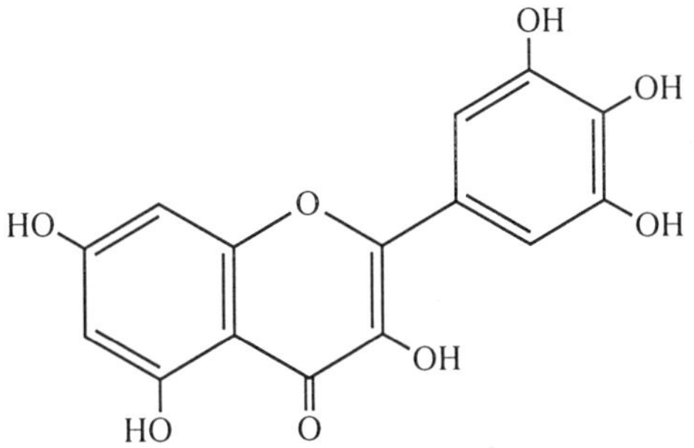

(5-26)

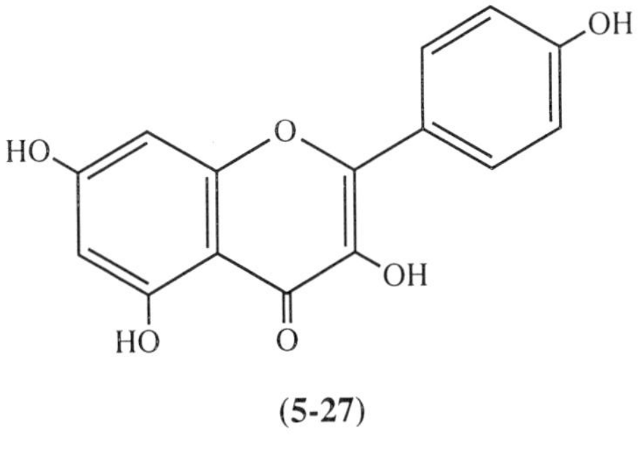

(5-27)

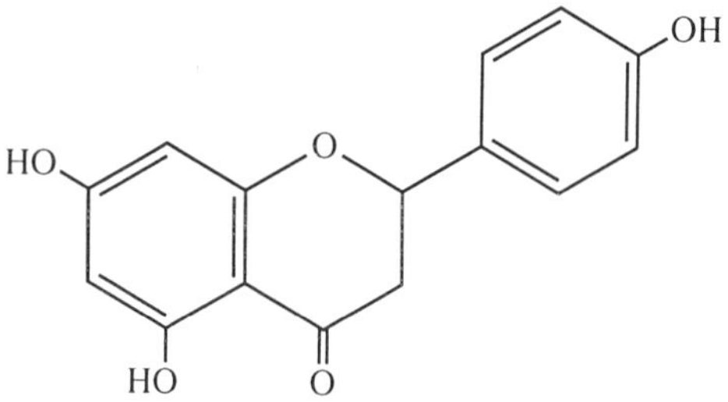

(5-28)

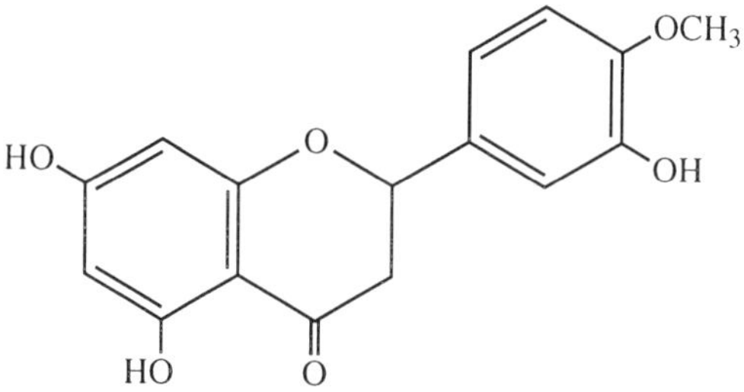

(5-29)

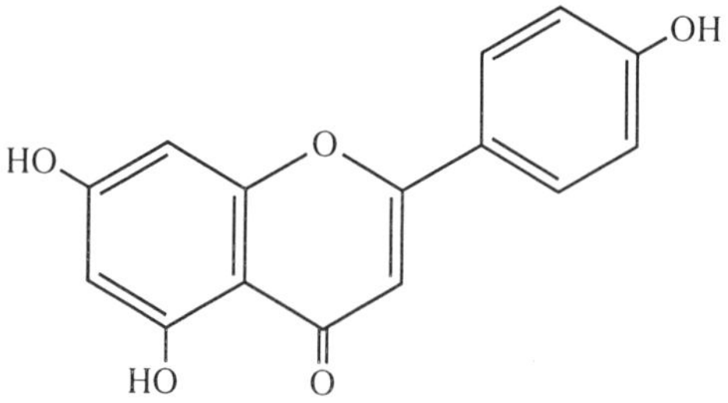

(5-30)

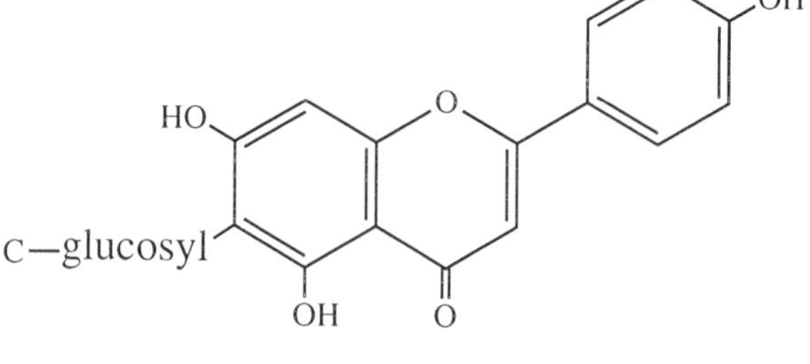

(5-31)

The reactions of flavonoid derivatives with superoxide, which largely seem to be oxidations, have been investigated by several groups. Cotelle et al. (1992), for example, synthesized ten substituted flavones that contained methoxyl or hydroxyl groups at various positions and exposed them to xanthine oxidase-generated $\cdot O_2^-$, with the result that some of them (especially those with pyrogallol substitution patterns such as 5-32) reacted directly with superoxide; intermediate free radicals were observed by ESR. It was hypothesized that these flavones were oxidized to o-quinones by two moles of $\cdot O_2^-$, which was in turn reduced to H_2O_2. Other flavones, which did not form detectable free-radical intermediates, appeared to act by inhibiting the enzyme used to generate superoxide.

(5-32)

Reactions of flavonoids with singlet oxygen have been briefly discussed. Flavonols such as quercetin (5-2) and fisetin (5-8), which quench singlet oxygen by chemical reaction, were generally more reactive (second-order quenching rate constants about 10^6 l/mol sec) than those of other types such as flavones. An exception was the highly efficient quencher, catechin (5-7; rate constant about 6×10^6 l/mol sec), which was a physical quencher that did not react with $^1O_2$ (Tournaire et al., 1993).

Anthocyanins are cationic polyphenols normally considered to be a class of flavonoids. They usually occur as glycosides and absorb wavelengths in the visible region of the spectrum. As water-soluble red or blue floral pigments, they have attracted a great deal of study from plant physiologists. They certainly serve to attract pollinators but it is also likely that they have other physiological functions. Several research groups have found indications of antioxidant activity for certain compounds of this series.

Igarashi et al. (1993) demonstrated that nasunin (5-33), malvin (5-34), and some other anthocyanins and their aglycones were able to inhibit the lipoxygenase-initiated bleaching of β-carotene in linoleic acid-containing micellar (Tween 80) suspensions as well as the autooxidation of linoleic acid at 70°C. The concentrations tested ranged from 5×10^{-5} to 2×10^{-4} M. Although the authors did not address mechanistic aspects of the inhibition, a possible scenario for their activity would be the formation of salt-like complexes at the surface of the micelle between the anthocyanin cation and the fatty acid anion. These complexes would exhibit high concentrations of readily oxidized polyphenol moieties at the interface where, presumably, the lipoxygenase would be generating reactive oxidants to attack the lipids.

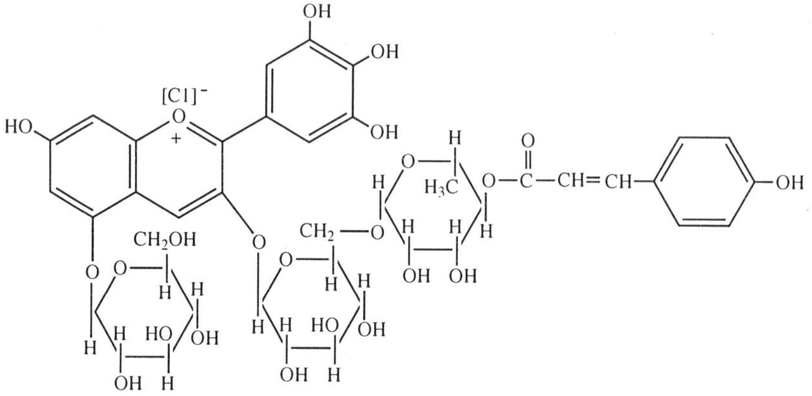

(5-33)

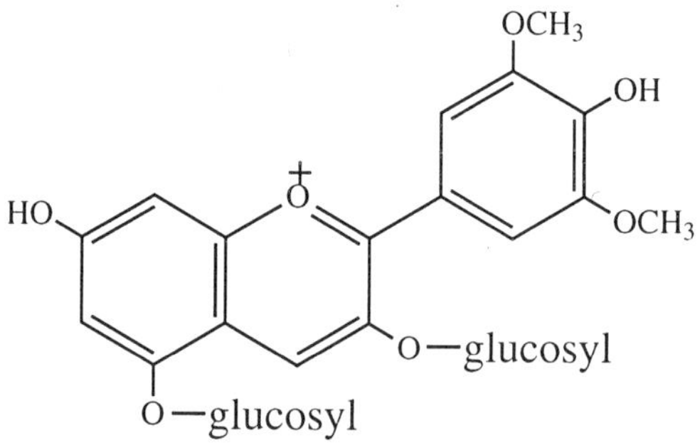

(5-34)

2. Isoflavonoids

Isoflavonoids (5-35) are of restricted distribution in the plant kingdom relative to the much more typical flavonoids; in fact, only one plant family, the Leguminosae, commonly contains them. In isoflavonoids, the position of the B ring is at the 3-, rather than the 2-position of the C ring. This fact precludes the occurrence of a hydrogen-bonded hydroxyl group at this position and therefore diminishes the probability of significant contributions by such a group to the antioxidant activity of an isoflavonoid. Furthermore, not many highly hydroxylated isoflavonoids are common in nature. Therefore, it is not surprising that few studies have shown these compounds to be particularly efficient antioxidants. Daidzein (5-36) and genistein (5-37), for example, were only marginally effective inhibitors of lard autooxidation at about 2mM (Dziedzic and Hudson, 1983a), and similar results have been reported in other systems (Pratt and Birac, 1979; Fleury et al., 1992). When used in food products in combination with other antioxidants, however, such as tocopherols or phospholipids, surprisingly high activity is sometimes observed (Hudson and Lewis, 1983).

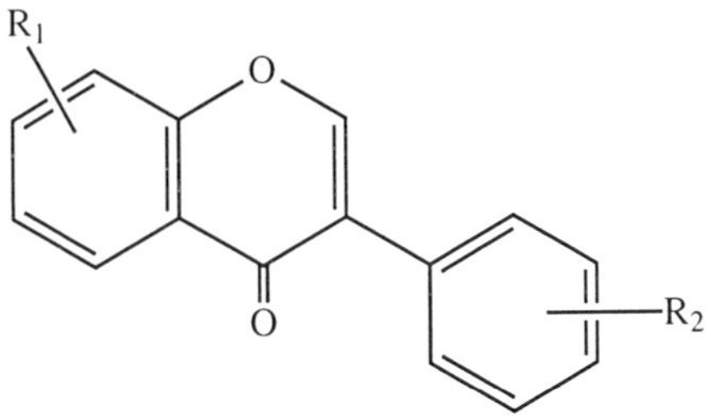

(5-35)

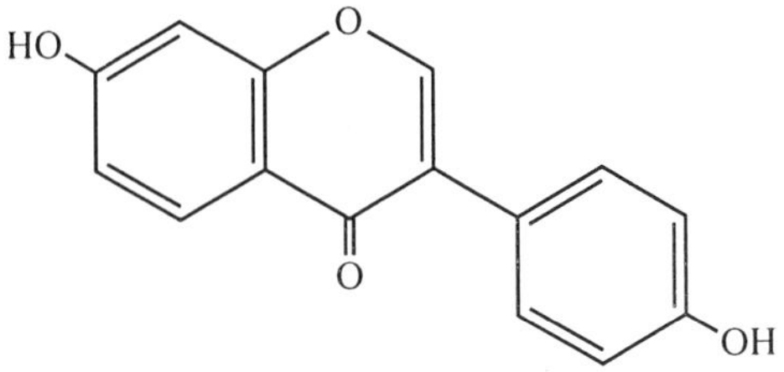

(5-36)

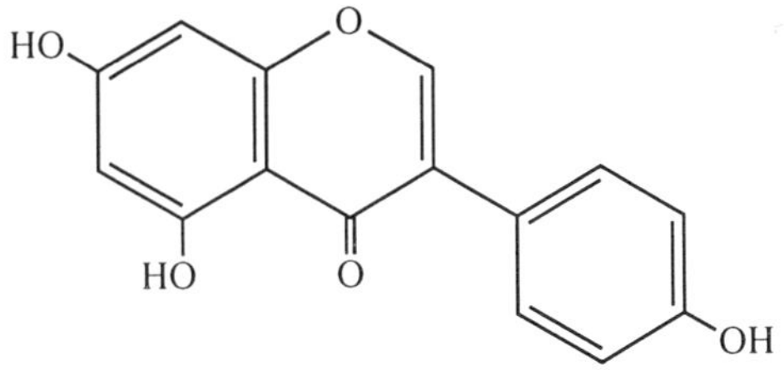

(5-37)

Recent studies on the reported anticarcinogenic activities of soybean phenolic compounds have rekindled interest in daidzein and genistein, which together with their glycosides are common in that plant. Genistein has been reported to inhibit the activities of a number of enzymes, and also to promote the synthesis of antioxidant enzymes such as catalase after dietary administration (Wei et al., 1995). It also may have as yet undefined specific antioxidant effects toward UV-induced peroxidation of DNA (Wei et al., 1996) or lipids (Record et al., 1995). It appears to be less effective at inhibiting iron-promoted oxidations, which would be consistent with the inferior metal ion-complexing ability in the C-ring of isoflavones relative to flavonols.

The isoflavonoid, 2″-O-glycosylisovitexin (5-38), isolated from green barley leaves, was shown to inhibit the formation of malondialdehyde from squalene or ethyl linoleate, induced either by UV irradiation or the Fenton reaction (Kitta et al., 1992). At concentrations of 1×10^{-4} M or above, 5-38 inhibited about 40% of MA formation under the conditions tested. It was shown to be less effective at pH 3.5 vs. 7.4, probably due to partial ionization of one or more phenolic hydroxy groups at the higher pH.

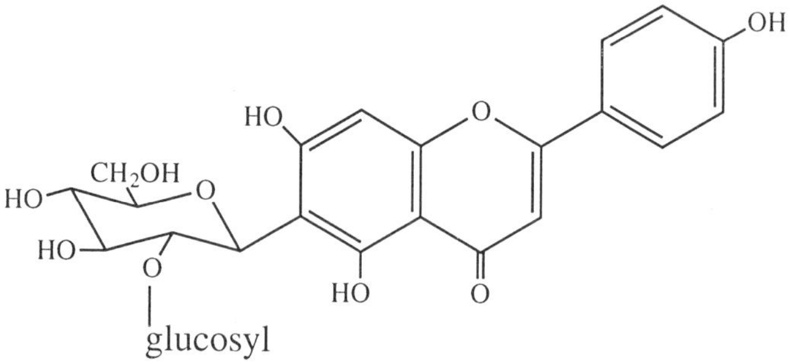

(5-38)

3. Chalcones and Catechins

Chalcones, which are natural polyphenolic precursors of flavonoids, possess many of their structural features and, as expected, have shown antioxidant activity in several investigations. In analogy to flavonoids, multiple or o-hydroxylation can lead to high activity, as in the case of butein (5-39) (Dziedzic and Hudson, 1983b). In some food products, butein has shown surprisingly high activity, being more effective at preventing oxidative lard decomposition than quercetin, α-tocopherol, or BHT (Namiki, 1990).

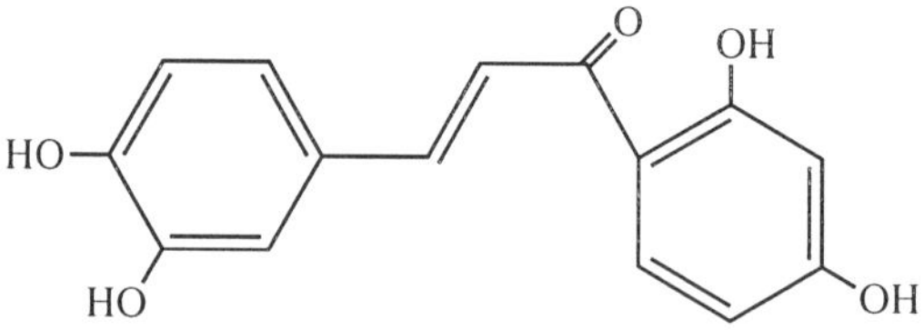

(5-39)

Catechins, or flavans (5-23), tricyclic polyphenols related to flavonoids and condensed tannins, have received much publicity recently (especially those of unfermented or "green" tea, which have been afforded a great deal of attention by the popular media) as potential anticancer agents. Green tea contains these compounds at levels (up to 18% of dry weight!) 3–6× greater than does fermented ("black") tea (Lunder, 1992). They do display antioxidant activity in some test systems (Matsuzaki and Hara, 1985: Hirose et al., 1990; Lunder, 1992). Catechin itself (5-7), as already mentioned, is a potent physical quencher for $^1O_2$ (Tournaire et al., 1993). Some reported structure–activity relationships were somewhat unusual, in that the more highly hydroxylated compounds were not the best antioxidants; seemingly, compounds with a dihydroxylated ring B were more potent than those bearing three hydroxyl groups in this ring. Perhaps this is related to the apparent ease of cleavage of this ring in the presence of peroxyl radicals, as shown by the isolation of an acrylic acid derivative, 5-40, as an oxidation product from (+)-catechin in about 50% yield (Hirose et al., 1990).

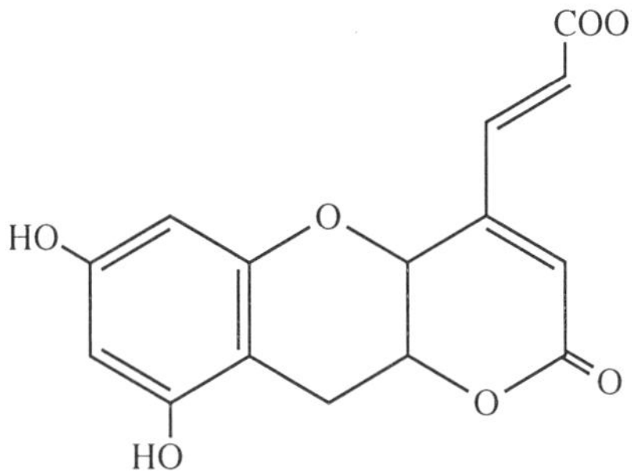

(5-40)

Catechins and epicatechin (5-41) have also been tested for their reactivity with oxidizing free radicals (Scott et al., 1993). They were found to be similar to some other substituted phenolic compounds (such as ferulic acid) insofar as their reactivities with HO· and Cl_3COO· were concerned, but they were not nearly as reactive as some other flavonoids (e.g., quercetin) or vitamin E analogues such as Trolox.

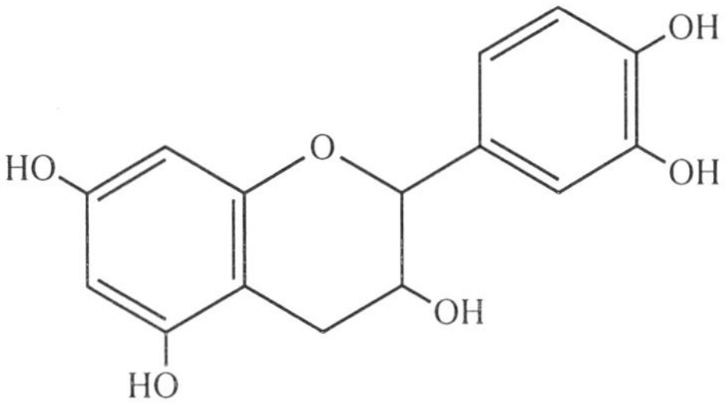

(5-41)

The oxidation products of catechins are a heterogeneous group of phenol coupling products, exemplified by theaflavin (5-42). They accumulate in black tea leaves and are extracted into the leaf infusions, and also display rather high antioxidant activity in some systems. For example, theaflavin had a higher antiperoxidative activity (as measured by the TBA test) than BHA, BHT, or α-tocopherol in rat liver homogenates (Yoshino et al., 1994). A concentration of only 8.7×10^{-6} M was sufficient to inhibit 50% of the oxidation under the test conditions. These authors also reported that a crude black tea extract had about the same antioxidative effectiveness as a green tea extract.

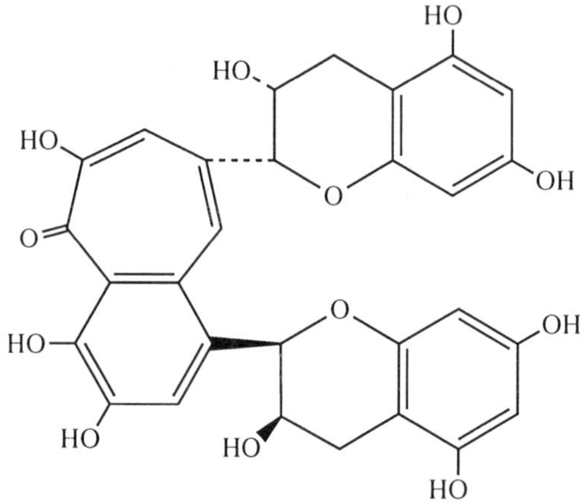

(5-42)

D. Phenolic Acids and Derivatives

1. Free Phenolic Acids

Metabolites of the shikimic acid pathway, and in particular compounds derived from the C_6–C_3 phenylpropanoid unit, are virtually universal in plant tissues, and are especially abundant in seeds and barks. The basic structural unit undergoes many alterations in the biosynthesis of phenylalanine, tyrosine, tannins, flavonoids, lignan, and lignin. Many of these compounds have been reported to be allelochemicals and structural building blocks. In plants, free phenolic acids principally occur as substituted benzoic (C_6–C_1: 5-43) and cinnamic (C_6–C_3: 5-44) types. Most tests of the antioxidant effectiveness of

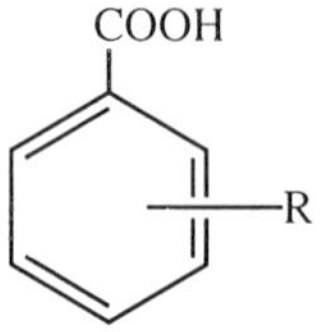

(5-43)

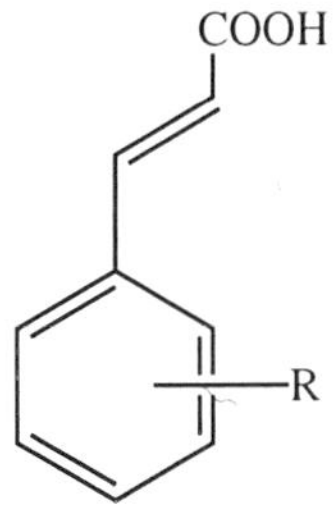

PHENOLIC AND ENOLIC ANTIOXIDANTS 111

(5-44)

these compounds have shown that the cinnamic derivatives are superior, probably because of the stability provided to intermediate free-radical forms by the extended conjugation in the side chain (see, e.g., Marinova and Yanishlieva, 1992). However, in most cases, their efficiencies are not particularly high, and rather high concentrations of the compounds are usually required to demonstrate significant protection.

Most reports have indicated that both benzoic acid and cinnamic acid derivatives vary widely in their antioxidant potential depending on the hydroxylation pattern on the benzene ring. It would be anticipated that increasing the number or strength of electron-donating substituents would make a phenolic acid more readily oxidizable and therefore a better antioxidant. A further factor that would contribute to the efficacy of polyhydroxylated phenolic acids would be the expected stabilization of free radicals derived from them due to resonance structures such as 5-45.

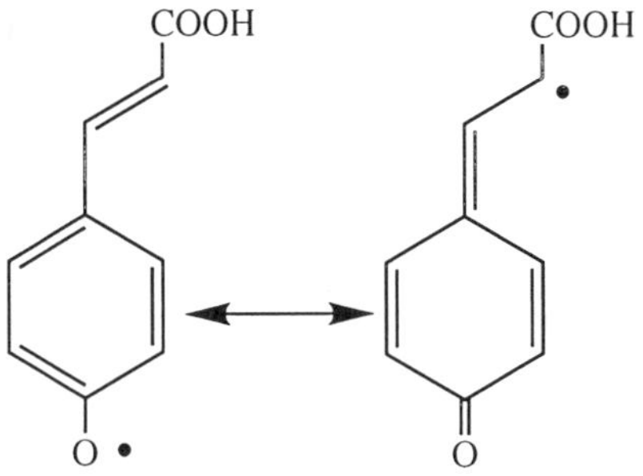

(5-45)

As expected, cinnamic acid and its monohydroxylated derivatives, in most tests, have meager antioxidant activity at best, whereas compounds with two hydroxyl or alkoxyl substituents possess some, and those with three can be quite active. An additional measure of activity is often observed in compounds having two or more adjacent phenolic OH groups, such as caffeic acid (5-46). The mechanistic rationales for the increased activity are not fully understood, although some authors suggest that this structural feature may be able to complex and inactivate certain transition metal ions. Quantitative evidence for this postulate, however, is largely lacking. It has been suggested that caffeic acid is an antioxidant only when iron(III) in the free form is absent from the medium; if iron is entirely in a bound form, for example as cytochrome c, caffeic acid appears to protect DNA from damage (Nakayama et al., 1993).

(5-46)

In any case, in most test systems caffeic acid usually has relatively high activity compared to other cinnamic and hydroxybenzoic acid derivatives. The combination of the cinnamoyl chromophore and the catechol moiety on the aromatic ring appear to lead to a highly stable and readily oxidized substance. When present in an oxidizing system, it quite often acts as an effective scavenger for a variety of free radicals. One example, given by Marinova and Yanishlieva (1992), is that it prolonged the lag time for lard oxidation at 100°C to >120 hr when added at a (somewhat high) concentration of 3 mM. It was at least six times as effective at this concentration than were seven other hydroxylated benzoic and cinnamic acid derivatives tested.

The kinetics of reaction of caffeic acid and a few related compounds as chain-breaking antioxidants were examined by Terao et al. (1993). A fairly standard lag-period analysis of the azo-initiated peroxidation of methyl linoleate in the solvent hexane-isopropanol-ethanol (8:3:0.1) was performed. Under these conditions, where the concentration of the antioxidant was 8.3 × 10^{-4} M and the reaction temperature was 37°C, caffeic acid was only 1/25 as active as vitamin E and was about 1/3 as active as BHA. Using this method, it was found to have activity comparable to that of a hydroxymethoxy cinnamic acid, ferulic acid, but much greater than that of a monohydroxy derivative, p-coumaric acid. Some of the reaction products formed when caffeic acid oxidizes include coumarins (Kagan, 1966; Larson and Rockwell, 1980) and lignan-like dimers that are substituted tetrahydrofuran dicarboxylic acids (5-47; Fulcrand et al., 1994). Presumably, all of these products arise by way of phenoxy (semiquinone) radical and o-quinone intermediates.

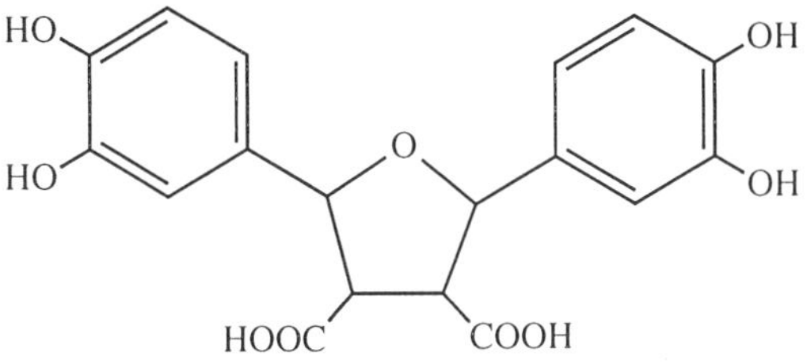

(5-47)

The antioxidant properties of ferulic acid (5-48) have been reviewed by Graf (1992). It has been used in some food products and sunscreen formulations. The compound normally exists in thermal equilibrium of about a 3:1 E:Z (*trans:cis*) ratio. Ultraviolet spectra of the two forms differ, with the Z isomer having an absorbance maximum at 316 nm, in the UV-B region, whereas the E form has two maxima of about equal intensity at 284 and 309 nm. In either case, light absorption in this region of the spectrum produces an excited state that is able to undergo E:Z isomerization and, possibly, engage in other free radical reactions (Fenton et al., 1978). Some antioxidant protection may be provided merely by UV absorption and the transfer of the potentially damaging radiation into "meaningless" thermal energy driving the isomerization equilibrium.

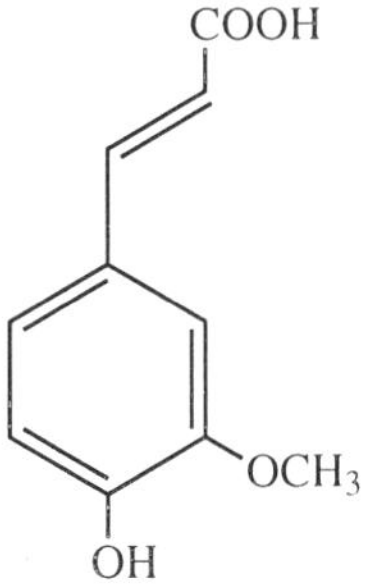

(5-48)

In most studies with ferulic acid, the necessary concentrations required to demonstrate antioxidant activity have been well above 1 mM. However, in a kinetic analysis of the azo-initiated peroxidation of methyl linoleate in the solvent hexane-isopropanol-ethanol (8:3:0.1), the compound was used at a concentration of 8.3×10^{-4} M. The reaction temperature was 37°C. Under these conditions, ferulic acid was 1/29 as active as vitamin E. Its activity was fairly close to that of caffeic acid (Terao et al., 1993).

The reactions of ferulic acid with peroxyl radicals have not been extensively studied. With trichloromethylperoxyl radicals, which are extremely reactive, a rate constant of 5.6×10^6 l/mol sec was determined by pulse radiolysis (Scott et al., 1993). This value was roughly an order of magnitude lower than that determined for the flavonoid, quercetin, and over thirty times lower than that for the vitamin E analog, Trolox (see Table 5-1). This is consistent with the order of antioxidant activity observed in most studies with these compound types.

The structure of sinapic acid (5-49) would suggest that it should have considerable antioxidant ability, and it has been found to be one of the active phenolic antioxidants in mustard and rape (Kozlowska et al., 1990). It inhibits autooxidation of fatty acid esters in a sunflower oil system (Yanishlieva and Marino, 1995). Ferulic acid was also active in this system; both seemed to function as chain-breaking antioxidants. Sinapic acid, like other cinnamates, undergoes photochemical E:Z isomerization and does so at a very high rate (Fenton et al., 1978).

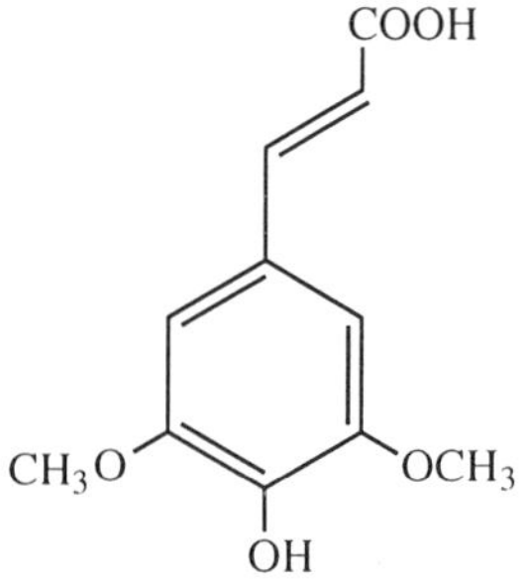

(5-49)

Among naturally occurring substituted benzoic acids, gallic acid (5-50) with its three hydroxyl groups is probably the most readily oxidized and would be expected to be the best antioxidant. As a constituent of gallotannins (esterified to glucose), it presumably contributes significantly to the antioxidant activity of such compounds, as well. Gallic acid reacts at a rather slow and insignificant rate with superoxide, but does react with a peroxyl radical ($Cl_3COO\cdot$) with a second-order rate constant in the 10^5 l/mol sec range

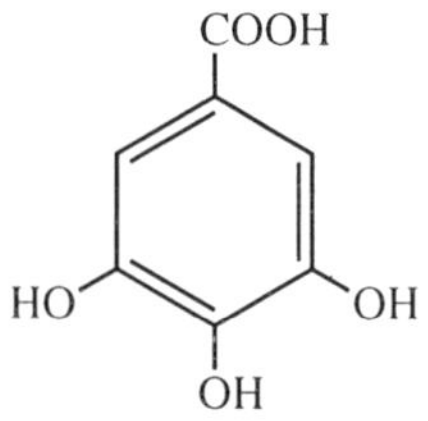

(5-50)

(Aruoma, 1993; Aruoma et al., 1993). However, in the latter regard it was (for some reason) tens or hundreds of times less reactive than its methyl, propyl, or lauryl esters.

In addition, gallic acid and related phenols in red wines have been suggested to be responsible for the "French paradox," that is, the fact that residents of France have lower rates of cardiovascular disease than those of other countries, despite consuming a diet high in fats. The antioxidant activity of red wines was, at least in one study of lipoprotein autooxidation (Frankel et al., 1995), more than tenfold higher than those of white wines. The activity showed strongest correlation ($r = 0.92$) with the gallic acid content of the wine.

Salicylic acid (5-51) has also been suggested to contribute to the antioxidant properties of red wines and the "French paradox." The reactivity of salicylic acid with oxidants is low relative to more highly substituted benzoic acids, but because of its structure it also has strong metal-complexing ability. It has been suggested that the antioxidant activity of salicylic acid in cells is due to the formation of iron-containing salicylate complexes that have reduced activity to participate in the Fenton reaction, releasing hydroxyl radicals (Cheng et al., 1996). Consistent with the hypothesis, salicylate was able to prevent the iron salt-catalyzed H_2O_2 oxidation of calf thymus DNA, but not that promoted by copper ion, which was not deactivated by salicylate.

(5-51)

2. Phenolic Esters and Amides; Tannins

Phenolic acids occur in a variety of derivatized forms, especially in plant tissues. Some of these compounds are of great interest because of their physiological and/or biochemical properties. For example, capsaicin (5-52), the active principle conferring the fiery taste of red pepper, is an amide containing a vanillic acid moiety. Esters of some phenolic acids with sugars or other carbohydrates are also common; they also have interesting physiological

(5-52)

properties as growth inhibitors or defensive substances. For the purposes of this chapter, however, we will largely focus on the functions of these compounds as antioxidants. The antioxidant activity of most of them depends, as nearly as can be determined, almost entirely on the phenolic acid portion of the structure; however, the distribution of the compound in the cell may be strongly influenced by the alcohol moiety.

Ferulic acid largely exists in the form of esters with carbohydrates, including carbohydrate polymers that make up part of the structural tissue of the cell wall. However, other esters that are oil-soluble, such as sterol ferulates, occur in some plants. For example, cycloartenyl ferulate (5-53) and related compounds occur in rice bran oil. These compounds have been demonstrated to be lipid peroxidation inhibitors, although to demonstrate much effect the concentration had to be about 50 mM, and the efficiency was not much different from that of ferulic acid itself (Yagi and Ohishi, 1979). Black pepper (*Piper nigrum*) contains several phenolic amide derivatives, including a derivative of ferulic acid (5-48) that was reported to be a better antioxidant than vitamin E (Nakatani et al., 1986).

(5-53)

Caffeic acid esters are rather widespread in nature. For example, caffeoyl quinate (5-54), also known as chlorogenic acid, is found in numerous plants including coffee, potatoes, and other commonly used food plants as well as the leaves, fruits, and other tissues of many dicotyledonous species. Chlorogenic acid exhibits relatively high antioxidant activity due to its parent phenolic acid, but in some test systems the quinate moiety also seems to contribute, making it more effective than the free acid (Pratt and Hudson, 1990). Rosmaric acid (5-55), another ester of caffeic acid, occurs in rosemary (*Rosmarinus officinalis*), a commonly used spice. It was reported to have antioxidant activity comparable to that of the free acid (Schuler, 1990). Benzyl caffeate (5-56), a closely related ester, has been isolated from propolis, a resinous material obtained from beehives (Yamauchi et al., 1992); it showed some effect in inhibiting thermal autooxidation of methyl linoleate, and may have been superior in activity to the flavonoids that were also present in the material. Sterol and triterpene alcohol esters of caffeic acid have been identified in canary seed (*Phalaris canadensis*) and tested as antioxidants for lard and other oils (Takagi and Iida, 1980).

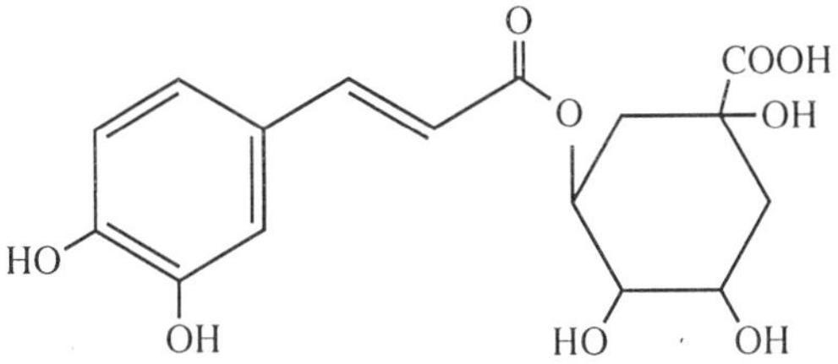

(5-54)

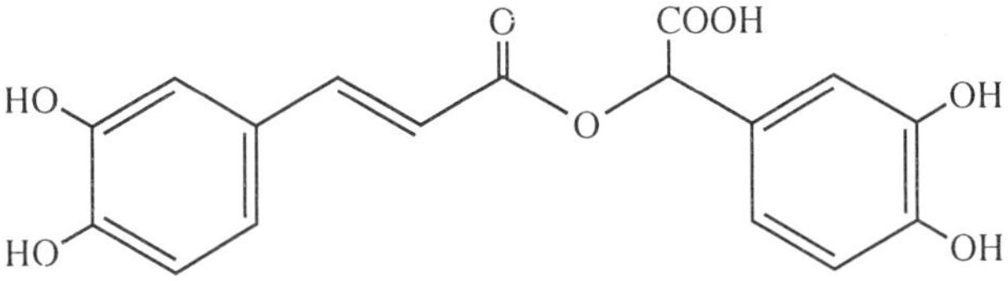

(5-55)

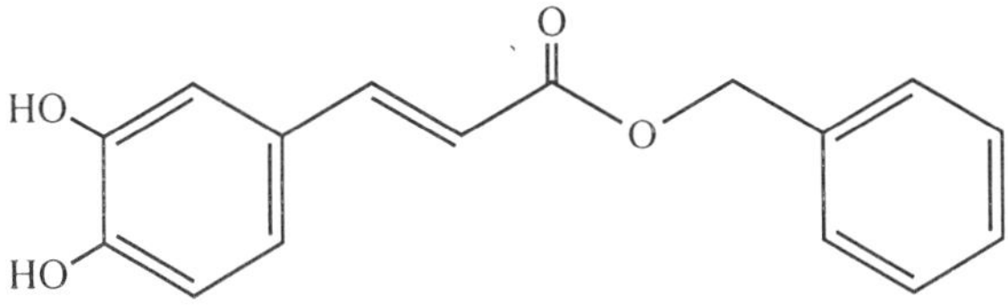

(5-56)

Gallic acid derivatives are often powerful antioxidants, as is shown by the use of the simple ester, propyl gallate (5-57), as one of the more common synthetic antioxidants for foods and other products. (The octyl and dodecyl esters are also used.) The compound is useful in foods because of its low volatility (and consequent lack of "phenolic" odor). Because propyl gallate forms colored complexes with metal ions, it is usually used in conjunction with a chelating agent such as citric acid. These complexes may contribute in some cases to the reported pro-oxidant activities of gallic acid and its derivatives (Aruoma et al., 1993).

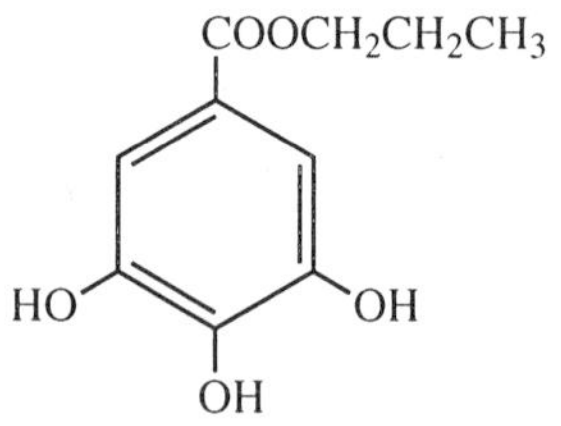

(5-57)

In nature, gallic acid is commonly esterified to the hydroxyl groups of a sugar, usually glucose, to form a series of common plant constituents referred to as gallotannins or hydrolyzable tannins. Many of these compounds have complex structures due to the phenol coupling of adjacent galloyl groups. A representative example (geraniin) is shown in 5-58. In addition, an ill-defined polymer, tannic acid, occurs as a mixture of esters containing between 5 and 12 galloyl units per mole of glucose.

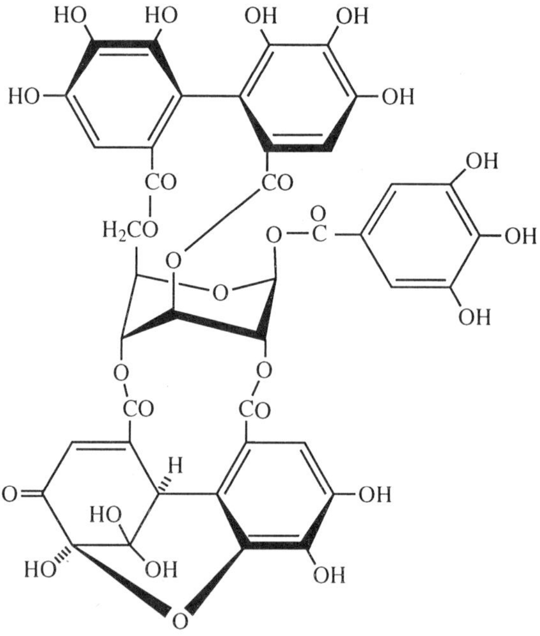

(5-58)

Several of these compounds have been demonstrated to be antioxidants or radical scavengers. The subject has been reviewed by Okuda et al. (1992). In one example, the copper(II)-catalyzed oxidation of ascorbic acid was shown to be inhibited by gallotannin derivatives such as tannic acid or geraniin, but only to a much lesser extent by gallic acid itself. Although some of the activity was attributed to metal ion complexation, the tannin was also shown to form one or more highly stable free radicals. Tannins also appear to be scavengers of singlet oxygen, hydroxyl radical, and superoxide radical (Masaki et al., 1994).

Ellagic acid (5-59) is an internal lactone derived from gallic acid and protocatechuic acid, found in some fruits and vegetables, that exhibits significant anticarcinogenic and antimutagenic activity. The compound has been reported to bind to DNA or to alter metabolic pathways, but its structure also suggests a potential for antioxidant activity. It has been shown to inhibit NADPH- or hematin-induced lipid peroxidation in mouse liver microsomes; a concentration of about 5×10^{-4} M was required to achieve 50% inhibition (Khanduja and Majid, 1993). It also reacts readily with diol epoxides of some mutagenic aromatic hydrocarbons (Sayer et al., 1982), suggesting that it might terminate oxidation processes by quenching other electrophiles. Okuda et al. (1992) reported that it was a strong inhibitor of the copper(II)-catalyzed autooxidation of ascorbic acid.

(5-59)

E. Lignans

C6–C3 dimers of varying degrees of complexity have been found to be antioxidants. As in other cases involving phenolic antioxidants, activity tends to be correlated with the number of phenoxyl or alkoxyl substituents in the compounds. For example, kadsurin (5-60), a compound isolated from a plant widely used in Chinese traditional medicine, *Kadsura heteroclita*, has six such substituents on its two benzene rings (Yang et al., 1992). The compound, when administered orally to mice that had been exposed to CCl_4, prevented lipid peroxidation in liver and serum. Related compounds, such as schisanhenol (5-61, also with six phenoxy and alkoxy substituents) from *Schisandra rubriflora*, were active inhibitors of iron-induced lipid peroxidation in rat liver microsomes (Lu and Liu, 1992).

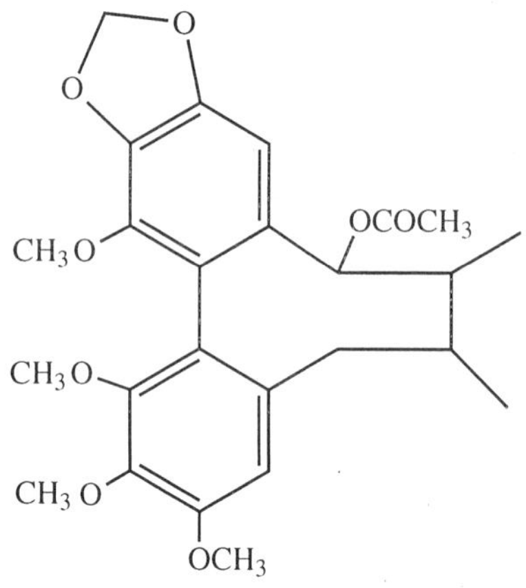

(5-60)

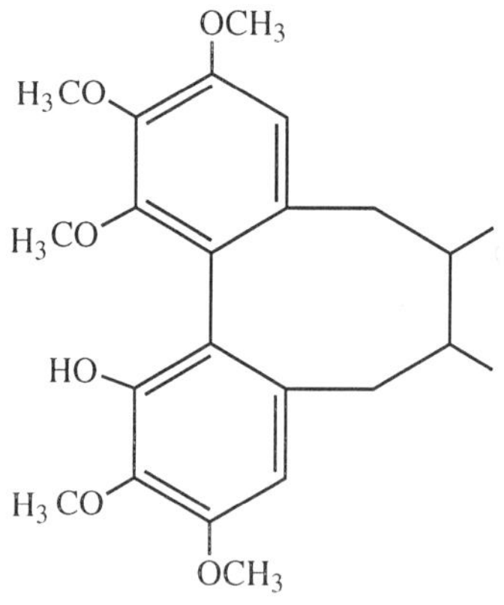

(5-61)

Although nothing is known with certainty about the *in vivo* mechanism of antioxidation by kadsurin and related compounds, intoxication with chlorinated methanes is known to involve reactive chloromethyl and chloromethylperoxyl radical intermediates.

Nordihydroguaiaretic acid (5-62) is a lignan-like C6–C3 dimer, occurs naturally as a major constituent of the surface resin of the creosote bush, *Larrea tridentata*. At one time, it was used as an antioxidant for foods, but has been withdrawn due to undesirable toxicological properties (Schuler, 1990). Not much is known concerning the mechanisms of its antioxidant activity, but it does appear to react quite readily with peroxyl radicals (Belyakov et al., 1995; Table 5-3).

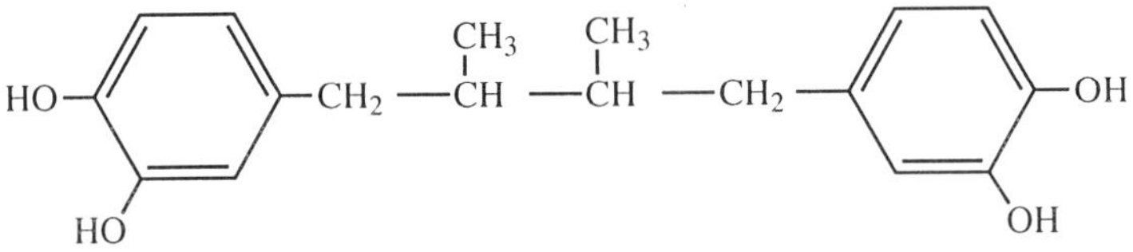

PHENOLIC AND ENOLIC ANTIOXIDANTS

(5-62)

F. Cucurmin and Derivatives

The rhizomes of tropical gingers and turmerics (e.g., *Zingiber* and *Curcuma* spp.) contain cucurmin (5-63), as well as less-oxygenated cucurmin derivatives 5-64 and 5-65. The antioxidant activity of these latter derivatives was greater than that of vitamin E, and practically identical to that of curcumin, in a Fe(II)-linoleic acid peroxidation test (Jitoe et al., 1992). Cucurmin is a very interesting substance because it generates phototoxic oxidizing species, including HO· and H_2O_2, when exposed to light (Tønnesen et al., 1993), but it also protects against lipid peroxidation as a radical scavenger (Tønnesen and Greenhill, 1992). The compound is a potent complexing agent for metal ions such as iron(III), which may help it play both of its roles as phototoxicant and protective agent. Cucurmin, like other β-diketones, exists in part in the enol form, which may also help to explain many of its unusual anti- and prooxidant properties.

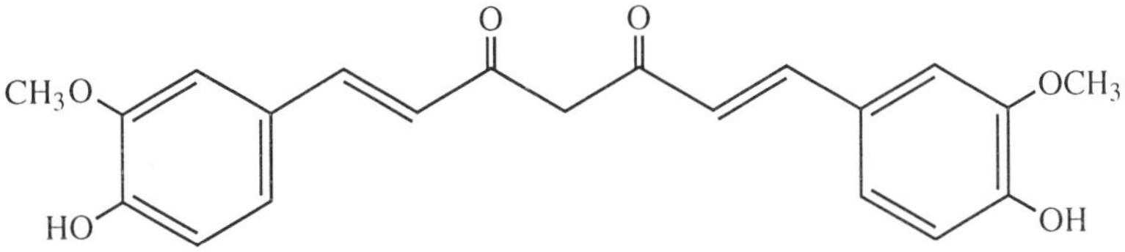

(5-63)

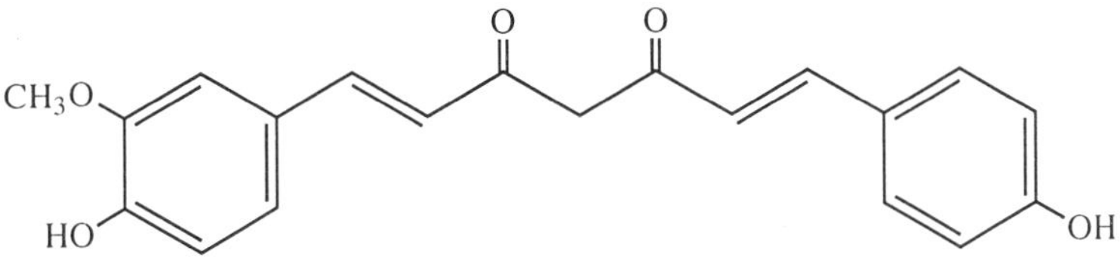

(5-64)

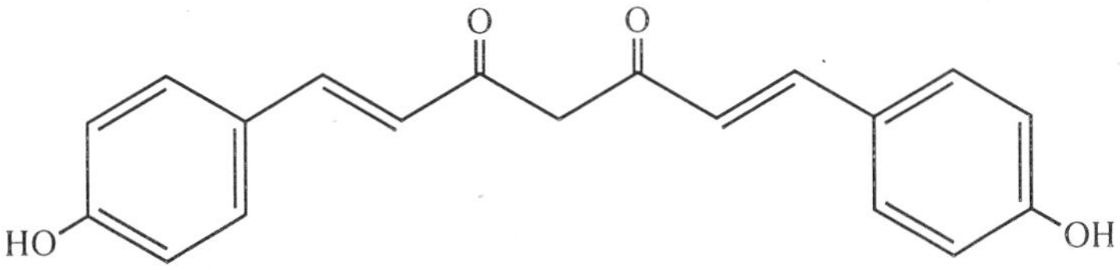

(5-65)

G. Hydroquinones and Quinones

1,4-Dihydroxybenzene derivatives (hydroquinones) are readily oxidized to 1,4-benzoquinones by many oxidants, including many one-electron oxidizing agents (Equation 5-8). An intermediate form, or semiquinone, is usually postulated as an intermediate in the redox process.

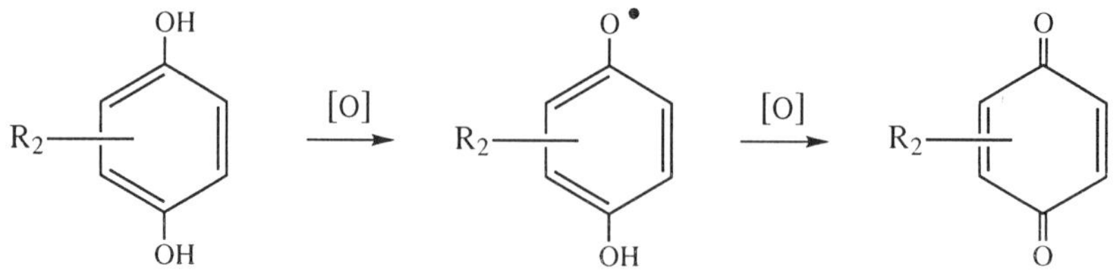

(Equation 5-8)

Several synthetic hydroquinones and quinones, such as hydroquinone itself and *t*-butyl hydroquinone, have been used as antioxidants for rubber and other polymers, as well as in foods. Quinones are known to act as vinyl polymerization inhibitors in synthetic polymers, probably because of their interactions with carbon-centered free radicals (cf. Chapter 4).

Perhaps the simplest naturally occurring hydroquinone derivative that has been shown to have antioxidant activity is arbutin (5-66), a glycoside of hydroquinone that is found in the above-ground parts of a number of medicinally useful or edible plants, including marjoram (*Origanum majorana*). The compound was found to be less active at inhibiting autooxidation of linoleate derivatives than either hydroquinone itself or vitamin E (Ioku et al., 1992), in either polar or nonpolar solvents (Figure 5-3).

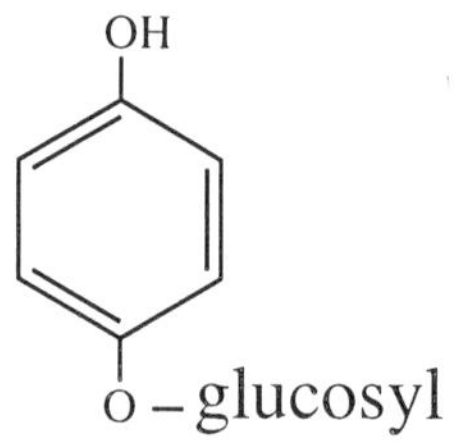

(5-66)

Ubiquinol-10 (5-67) occurs in human plasma and other tissues together with its oxidized form, ubiquinone. The hydroquinone form predominates in some organs such as heart, kidney, and liver, whereas the oxidized form is found in greater abundance in the brain and lung (Aberg et al., 1992). Ubiquinol has a structure suggestive of great antioxidant power, especially in lipid systems: two hydroxyl and two methoxyl groups plus an *o*-methyl substituent and

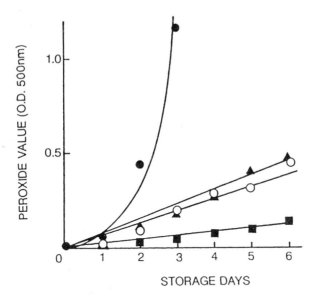

Figure 5-3 Effect of arbutin and other antioxidants on the autooxidation of linoleic acid in an aqueous ethanol solution. Solid circles, control; solid triangles, arbutin; open circles, vitamin E; solid squares, hydroquinone. Linoleic acid was present at 0.7 mM concentration in a mixture of 0.05 M phosphate buffer and ethanol. Additives were tested at equal weights (4 mg/test). (From Ioku, K. J., et al., 1992, *Biosci. Biotech. Biochem.*, 56, 1658–1659. Reprinted by permission.)

a lengthy side chain. Indeed, it is a quite good inhibitor of free radical reactions, if only with about 1/3 to 1/10 the activity of vitamin E in solution measurements (Yamamoto et al., 1990; Kagan et al., 1990; Mukai et al., 1993a). Another important feature of its *in vivo* activity its probably its potential to act as a hydrogen donor to vitamin E (Mukai et al., 1990; Ingold et al., 1993). It also is a potent singlet oxygen quencher (Mukai et al., 1993c). Ubiquinols and their derivatives exist in fairly high concentrations in many tissues and in mitochondria, and are therefore probably important contributors to radical scavenging in such tissues. It has even been suggested that ubiquinol-10 is a more significant antioxidant for membrane lipoproteins than vitamin E (Stocker et al., 1991).

(5-67)

There has been much less work done on the mechanisms of action of the oxidized forms of naturally occurring quinones, although several have been found to have antioxidant activity. For some time, for example, it has been known that both the oxidized and reduced forms of quinones such as ubiquinone (5-68) have antioxidant activity in some systems (Mellors and Tappel, 1966; Cabrini et al., 1986), although other authors (Lambelet et al., 1992) have claimed that it does not react with peroxidizing lipids, and that any antioxidant activity observed when the compound is present is due to its reduction to the semiquinone or hydroquinone form or to the presence of additional reductants. Other quinones and anthrones (compounds of the type 5-69) were reported to inhibit hydroperoxide-induced lipid peroxidation in rat hepatocytes as well as in a linoleic acid-peroxidizing medium (Malterud et al., 1993). The more effective compounds, however, also contained phenolic hydroxyl groups, suggesting that alternative antioxidant mechanisms might have been operating.

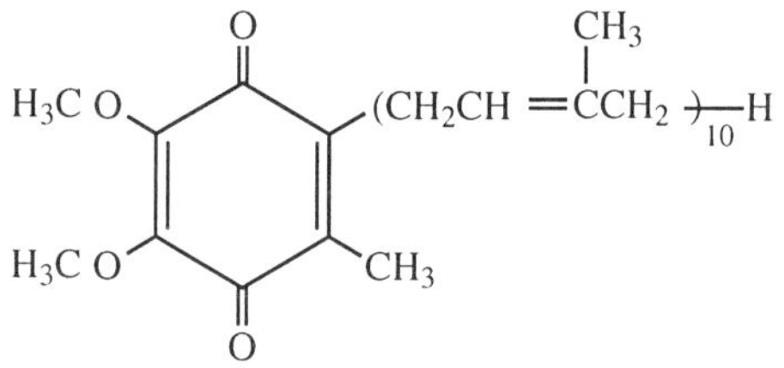

(5-68)

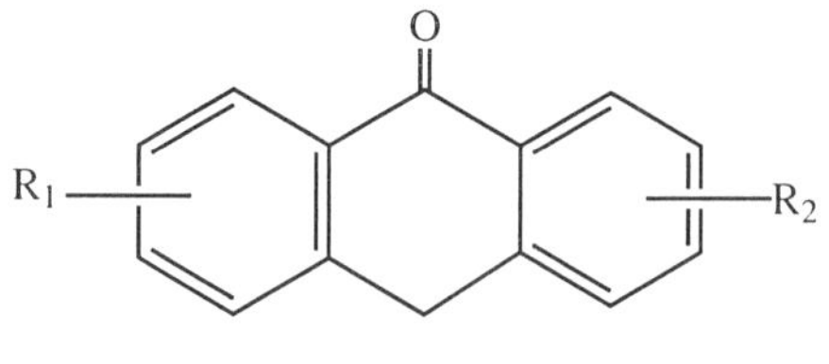

(5-69)

Quinones are known to react at very high rates with superoxide to form the semiquinone (Halliwell and Gutteridge, 1989), which might remove a potential source of potent oxidants. Reactions of ubiquinone and other biological quinones with other free radicals are not very well understood. Landi et al. (1990) have suggested that ubiquinone is able to react readily with peroxyl radicals derived from peroxidizing lipids, and Briviba and Sies (1994) reported that ubiquinone reacted with singlet oxygen at a rate roughly one-third of that of the hydroquinone form.

Several plants including rosemary (*Rosmarinus officinalis*) and tanshen (*Salvia miltiorrhiza*) contain modified diterpenoid *o*-quinones with antioxidant activity (at about 7×10^{-4} M) in a lard rancification assay. The rosemary quinones are oxidized forms of the catechols described below. Rosmariquinone (5-70) and tanshinone I (5-71) are typical representatives of this class of compounds.

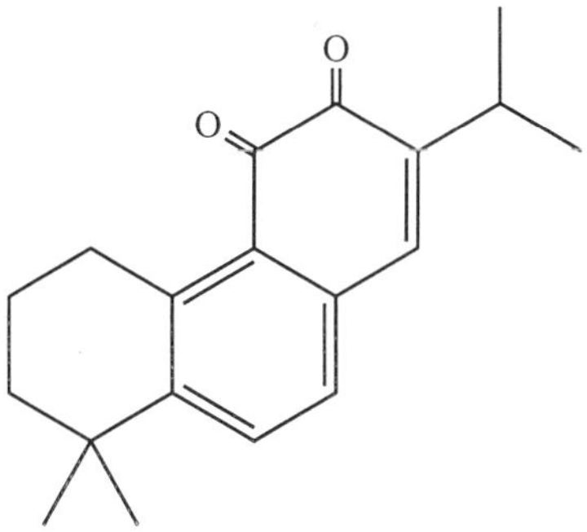

(5-70)

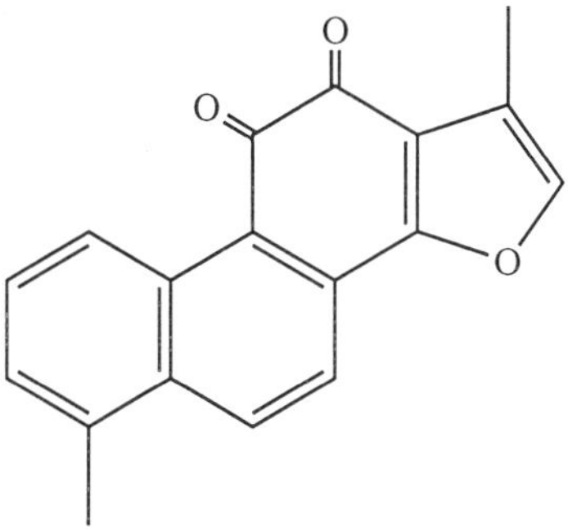

(5-71)

Rosmariquinone appears to have the highest activity of the group in this assay; its activity was comparable to that of the synthetic antioxidants, BHA and BHT (Wang and Gordon, 1992). However, even 9,10-phenanthrenedione (5-72) was an active inhibitor of the lipid oxidation in the test, suggesting that

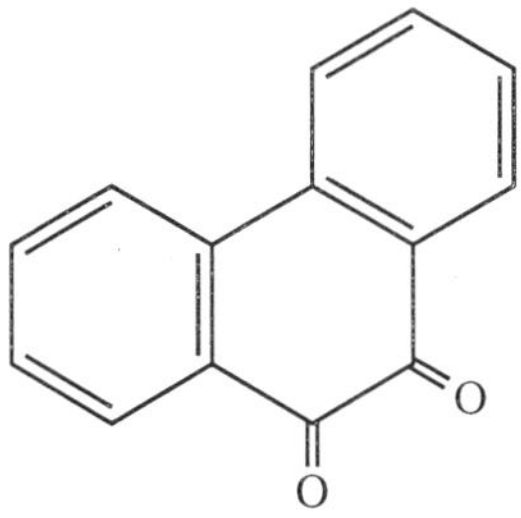

(5-72)

the activity was mainly due to the *o*-quinone functionality. The possibility that the quinone was acting as a complexing agent for trace metal ions was ruled out by adding iron salts to the mixture, which resulted in strong suppression of the observed inhibition.

Possibly, naturally occurring quinones could act in the same way in the presence of low oxygen concentrations. In other words, the quinone would have to compete with oxygen for R· radicals (see Chapter 4). However, there do not seem to have been any careful studies of this question, although support for an alkyl radical mechanism has come from a study of a reaction between methyl oleate and 9,10-phenanthrenedione (Weng and Gordon, 1992). When the reactants were heated together at 50°C for a prolonged period, an addition product (5-73) was isolated in unspecified yield. The mechanism proposed for the formation of the product was a radical-radical coupling between the semiquinone and an allyl radical of the ester.

$$\text{CH}_3(\text{CH}_2)_6-\overset{\text{HO}}{\underset{|}{\text{CH}}}\overset{\text{O}}{\underset{}{}}$$

$$\text{CH}_3(\text{CH}_2)_6-\text{CH}\diagdown\text{CH}$$
$$\parallel$$
$$\text{H}_3\text{COC}-(\text{CH}_2)_7\diagup\text{CH}$$
$$\parallel$$
$$\text{O}$$

(5-73)

H. Other Phenolic Compounds

A great variety of simple phenolic compounds occur widely in plants. Some are biosynthetically derived from terpenes, others by acetogenic mechanisms such as the shikimic acid pathway (Robinson, 1991). Quite a number of these compounds have been tested in one antioxidant assay or another, and many have shown activity, though the concentrations employed are sometimes unrealistically high.

Thymol (5-74) and carvacrol (5-75) are closely related monoterpenes found in the essential oils of many food plants such as thyme and marjoram. They have moderate activity as disinfectants and antiseptics. Both were found to significantly inhibit membrane lipid peroxidation at concentrations above 10^{-4} M (Aeschbach et al., 1994). In addition, they react with reasonably fast rates with the trichloromethylperoxyl radical (see Table 5-1), suggesting that they may be acting as chain-breaking antioxidants.

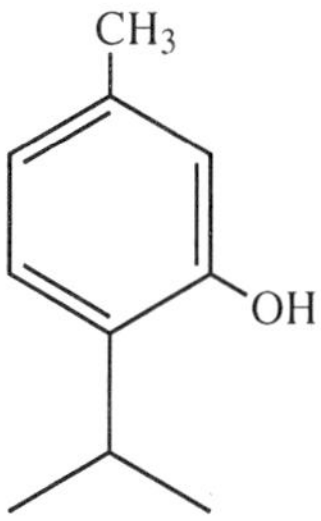

(5-74)

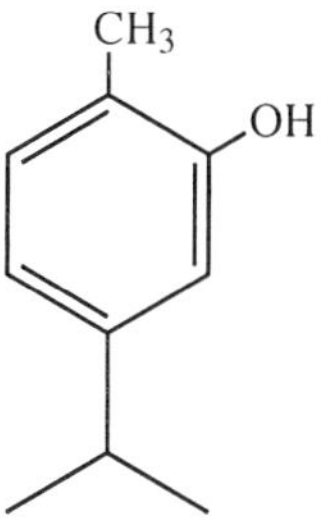

(5-75)

Purpurogallin (5-76) exists as a mixture of glycosides in plant galls (tumor-like growths, seen especially on the bark of trees, that are caused by insects, fungi, and other damaging agents). It is a deep red compound that is also formed (among many other products) by the chemical oxidation of gallic acid and of pyrogallol (1,2,3-trihydroxybenzene). It was patented as an antioxidant for vegetable and mineral oils and fuel products, but has also been shown to inhibit the peroxyl radical-induced hemolysis of erythrocytes (Sugiyama et al., 1993). In this system, it inhibited 50% of the observed lysis at a lower concentration (ca. 10^{-4} M) than was observed with the vitamin E analogue Trolox (ca. 6×10^{-4} M).

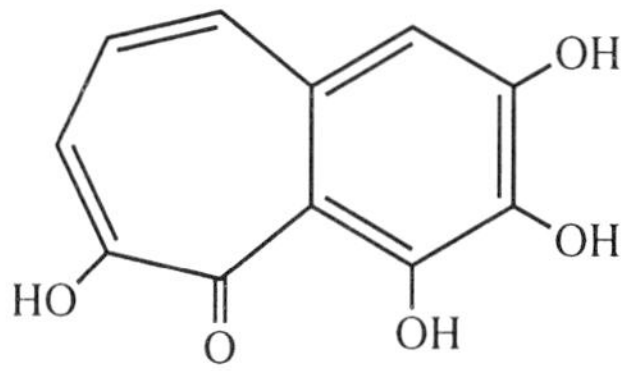

(5-76)

Olive oil contains a polar fraction that is enriched in phenolic acids and related compounds. Several of these substances show some antioxidant activity for the parent material. Among the most potent compounds contributing to the effect is hydroxytyrosol, or 3,4-dihydroxyphenylethanol (5-77). When added to refined olive oil at a concentration of 1.3 mM, it delayed the onset of peroxide formation (at 63°C) by a factor of 10–15 relative to a control (Papadopolous and Boskou, 1991). It was over twice as effective as BHA. The only compound of 12 tested that was superior was 3,4-dihydroxyphenylacetic acid.

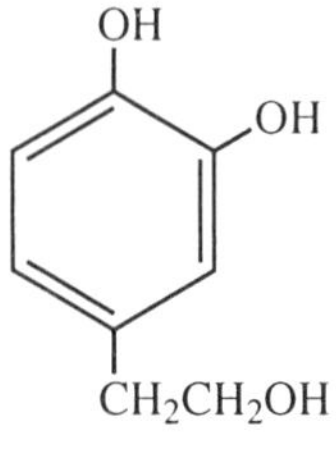

(5-77)

Sesame (*Sesamum indicum*) seed contains several active antioxidants, including the simple phenol sesamol (5-78), and some lignan-like compounds, sesamolin (5-79) and sesamolinol (5-80). The high antioxidant activity of sesamol is somewhat puzzling (Lyon, 1972). It is readily oxidized to a number of products, including dimers which also may possess antioxidant potential (Kikugawa et al., 1990).

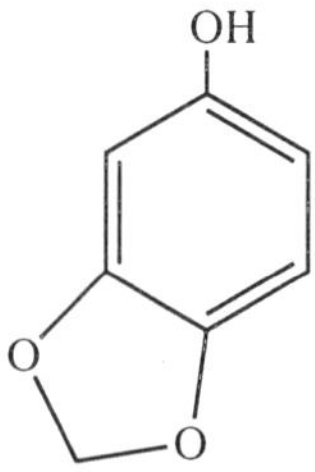

(5-78)

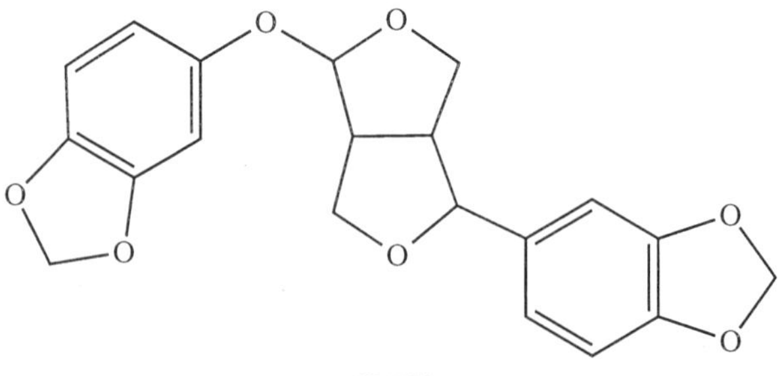

(5-79)

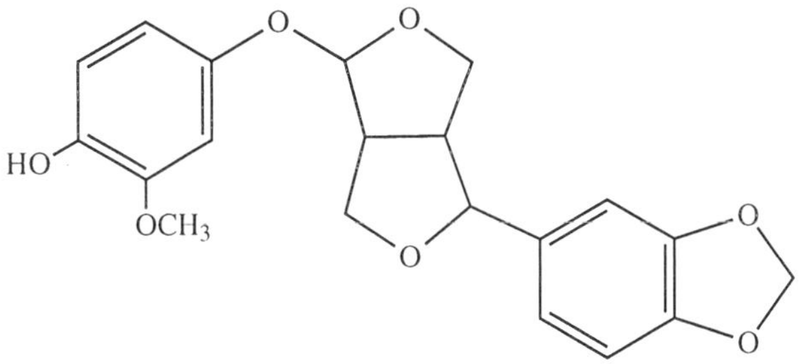

(5-80)

Long-chain aliphatic resorcinol derivatives have been identified in several plants. Struski and Kozubek (1992) tested a family of these compounds (5-81), isolated from rye grain, for their activity at inhibiting the peroxidation of phospholipid micelles. Although in these systems it was difficult to determine

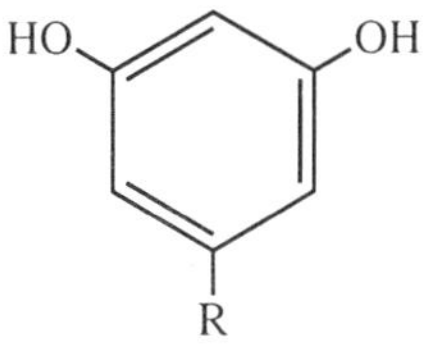

(R = unsatured C_{13} – C_{25} alkyl chains)

(5-81)

the absolute concentrations of the compounds in the peroxidizing system, significant inhibition of TBA-reactive substances was noted even at less than 5 mole % antioxidant relative to lipid (Figure 5-4). The structure of these compounds is reminiscent of vitamin E and it is likely that they would localize effectively in lipid bilayer portions of membranes.

A similar, but unusual, resorcinol derivative is resorstatin (5-82) from a bacterial species, *Pseudomonas* strain DC165. The compound showed activity

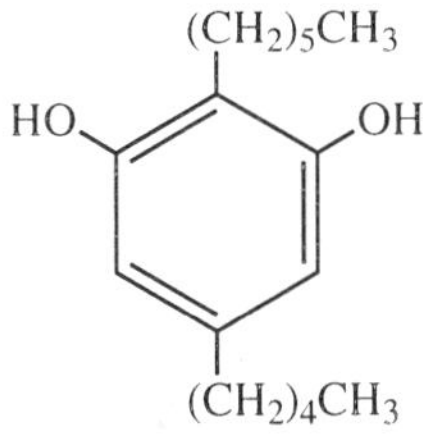

(5-82)

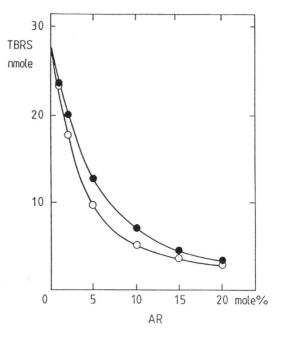

Figure 5-4 The effects of substituted resorcinols from rye grain on iron(II)-induced peroxidation of linoleic acid emulsions. AR = alkylresorcinol, TBRS = thiobarbituric acid-reactive substances. Solid circles, 5-n-heptadecylresorcinol; open circles, 5-n-heptadecenylresorcinol. (From Struski and Kozubek (1992). Reprinted by permission.)

in the inhibition of Fe(II)-ascorbate initiated lipid peroxidation in rat brain homogenates practically identical to that of BHT — virtually complete inhibition at concentrations >4 μM (Kato et al., 1993).

Rosemary (*Rosmarinus officinalis*) extracts display antioxidant activity, due at least in part to the quinones described above, but also to several related phenolic diterpenes: carnosol, carnosic acid (5-83 and 5-84), rosmanol,

(5-83)

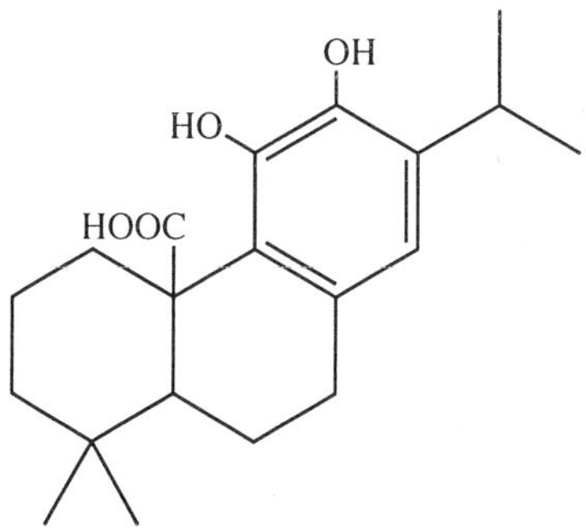

(5-84)

epirosmanol, and isorosmanol (5-85–5-87; Nakatani and Inatani, 1984). A method for separating these compounds by reverse-phase HPLC with electrochemical detection has been published (Schwartz and Ternes, 1992), but no attempt to correlate the antioxidative potency of the compounds with their half-wave potentials was made.

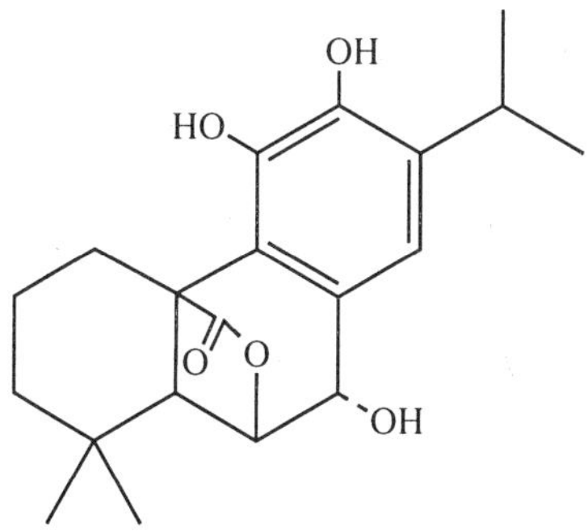

(5-85)

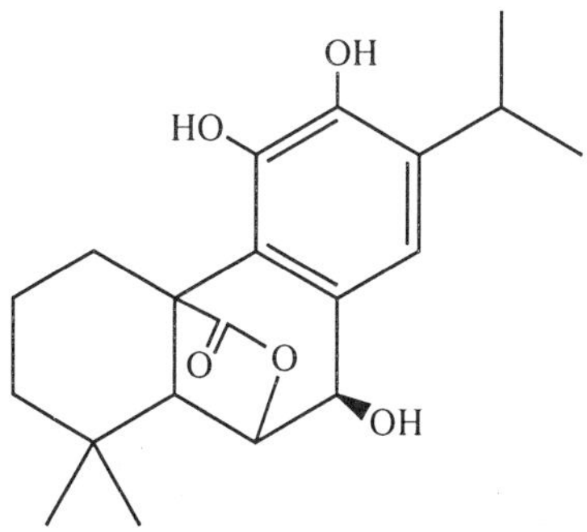

(5-86)

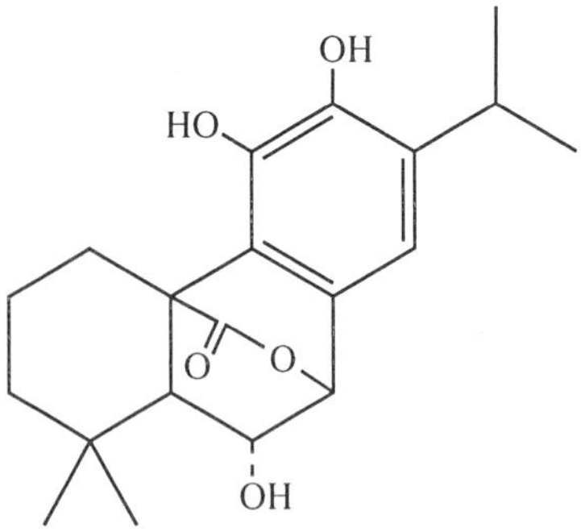

(5-87)

Many fungi contain unusual polyphenolic substances, and some of these show modest antioxidant activity. For example, atrovenetin (5-88), a derivative of the tricyclic ketone, phenalenone) was isolated from the mycelium of *Penicillium paraherquei* and tested for its ability to inhibit food lipid oxidation (Ishikawa et al., 1991). At a concentration of about 6×10^{-4} M, it increased the induction period for several vegetable oils from roughly 13 hr to more than 200 hr.

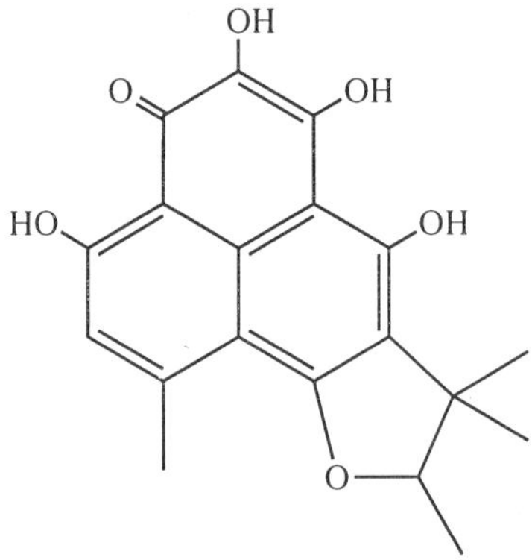

(5-88)

REFERENCES

Aberg, F., E. L. Appelkwist, G. Dallner, and L. Erstner. 1992. Distribution and redox state of ubiquinones in rat and human tissues. Arch. Biochem. Biophys. 295, 230–234.

Aeschbach, R., J. Loliger, B. C. Scott, A. Murcia, J. Butler, B. Halliwell, and O. I. Aruoma. 1994. Antioxidant actions of thymol, carvacrol, 6-gingerole, zingerone, and hydrosytyrosol. Food Chem. Toxicol. 32, 31–36.

Ariga, T. and M. Hamano. 1990. Radical scavenging action and its mode in procyanidins B-1 and B-3 from azuki beans to peroxyl radicals. Agric. Biol. Chem., 54, 2499–2504.

Aruoma, O. I. 1993. Free radicals and food. Chem. Brit. (March) 210–214.

Aruoma, O. I., A. Murcia, J. Butler, and B. Halliwell. 1993. Evaluation of the antioxidant and prooxidant actions of gallic acid and its derivatives. J. Agric. Food Chem. 41, 1880–1885.

Belyakov, V. A., V. A. Roginsky, and W. Bors. 1995. Rate constants for the reaction of peroxyl free radical with flavonoids and related compounds as determined by the kinetic chemiluminescence method. J. Chem. Soc. Perkin Trans. 2, 2319–2326.

Bendich, A., L. J. Machlin, O. Scandurra, G. W. Burton, and D. D. M. Wayner. 1986. The antioxidant role of vitamin C. Adv. Free Rad. Biol. Med., 2, 419–444.

Bielski, B. H. J. 1983. Evaluation of the reactivities of perhydroxyl radical/superoxide radical with compounds of biological interest. In G. Cohen and R. A. Greenwald, eds., *Oxyradicals and Their Scavenger Systems*. Vol. 1, Molecular aspects. Elsevier Biomedical, New York. pp. 1–7.

Bindoli, A., M. Valente, and L. Cavalli. 1985. Inhibition of lipid peroxidation by α-tocopherolquinone and α-tocopherolhydroquinone. Biochem. Int. 10, 753–761.

Bolland, J. L. and P. Ten Have. 1947. Kinetic studies in the chemistry of rubber and related materials. V. Inhibitory effect of phenolic compounds on the thermal oxidation of ethyl linoleate. Disc. Faraday Soc. 2, 252–260.

Buettner, G. R. 1988. In the absence of catalytic metals ascorbate does not autoxidize at pH 7. Ascorbate as a test for catalytic metals. J. Biochem. Biophys. Meth., 16, 27–40.

Burton, G. W., L. Hughes, and K. U. Ingold. 1983. Antioxidant activity of phenols related to vitamin E. Are there chain-breaking antioxidants better than α-tocopherol? J. Am. Chem. Soc. 105, 5950–5951.

Buxton, G. V., C. L. Greenstock, W. P. Helman, and A. B. Ross. 1988. Critical review of rate constants for reactions of hydrated electrons, hydrogen atoms, and hydroxyl radicals in aqueous solution. J. Phys. Chem. Ref. Data 17 514–886.

Cabelli, D. E. and B. H. J. Bielski. 1983. Kinetics and mechanism for the oxidation of ascorbic acid/ascorbate by $HO_2/\cdot O_2^-$ radicals. A pulse radiolysis and stopped-flow photolysis study. J. Phys. Chem., 87, 1809–1817.

Cadenas, E. 1995. Mechanisms of oxygen activation and reactive oxygen species detoxification. In S. Ahmad, ed., *Oxidative Stress and Antioxidant Defenses in Biology*. Chapman and Hall, New York. Chapter 1, pp. 1–61.

Cheng, I. F., C. P. Zhao, A. Amolins, M. Galazka, and L. Doneski. 1996. A hypothesis for the *in vivo* antioxidant action of salicylic acid. Biometals 9, 285–290.

Cotelle, N., J. L. Bernier, J. P. Hénichart, J. P. Catteau, E. Gaydou, and J. C. Wallet. 1992. Scavenger and antioxidant properties of ten synthetic flavones. Free Rad. Biol. Med. 13, 211–219.

Dziedzic, S. J. and B. J. F. Hudson. 1983a. Hydroxyisoflavones as antioxidants for edible oils. Food Chem. 11, 161–166.

Dziedzic, S. J. and B. J. F. Hudson. 1983b. Polyhydroxy chalcones and flavanones as antioxidants for edible oils. Food Chem. 12, 205–212.

Fenton, T. W., M. M. Mueller, and D. R. Clandinin. 1978. Isomerization of some cinnamic acid derivatives. J. Chromatogr. 152, 517–522.

Fessenden, R. W. and N. C. Verma. 1978. A time-resolved electron spin resonance study of the oxidation of ascorbic acid by the hydroxyl radical. Biophys. J. 24, 93–101.

Fleury, Y., D. H. Welti, G. Phillipossian, and D. Magnoloto. 1992. Soybean (malonyl) isoflavones. Characterization and antioxidant properties. In M.-T. Huang, C.-T. Ho, and C. Y. Lee, eds., *Phenolic Compounds in Food and Their Effects on Health.* II. Antioxidants and cancer prevention, Am. Chem. Soc. Sympos. Ser. #507, Chapter 8, pp. 98–113.

Foti, M., K. U. Ingold, and J. Lusztyk. 1996. The surprisingly high reactivity of phenoxyl radicals. J. Am. Chem. Soc. 116, 9440–9447.

Frankel, E. N., A. L. Waterhouse, and P. L. Teissedre. 1995. Principal phenolic phytochemicals in selected California wines and their antioxidant activity in inhibiting oxidation of human low-density lipoproteins. J. Agric. Food Chem. 43, 890–894.

Fryer, M. J. 1992. The antioxidant effects of thylakoid vitamin E (α-tocopherol). Plant Cell Environ. 15, 381–392.

Fulcrand, H., A. Cheminat, R. Broillard, and V. Cheynier. 1994. Characterization of compounds obtained by chemical oxidation of caffeic acid in acidic conditions. Phytochemistry 35, 499–505.

Graf, E. 1992. Antioxidant potential of ferulic acid. Free Rad. Biol. Med. 13, 435–448.

Groden, D. and E. Beck. 1979. H_2O_2 destruction by ascorbate-dependent systems from chloroplasts. Biochim. Biophys. Acta 546, 426–453.

Halliwell, B. and J. M. C. Gutteridge. 1989. *Free Radicals in Medicine and Biology.* 2nd ed. Clarendon Press, Oxford, UK.

Hayashi, T., A. Kaneyoshi, M. Nakamura, M. Tamura, and H. Shirahama. 1992. Reduction of α-tocopherylquinone to α-tocopherylhydroquinone in rat hepatocytes. Biochem. Pharmacol. 44, 489–493.

Hirose, Y., H. Yamaoka, and M. Nakayama. 1990. Oxidation product of (+)-catechin from lipid peroxidation. Agric. Biol. Chem. 54, 567–569.

Hudson, B. J. F. and J. I. Lewis. 1983. Polyhydroxy flavonoid antioxidants for edible oils. Food Chem. 10, 111–120.

Igarashi, K., T. Yoshida, and E. Suzuki. 1993. Antioxidative activity of nasunin in chouja-nasu (little eggplant, *Solanum melongena* L. 'Chouja'). Nippon Shokyuhin Kogyo Gakkaishi 40, 138–143.

Ingold, K. U., V. W. Bowry, R. Stocker, and C. Walling. 1993. Autoxidation of lipids and antioxidation by α-tocopherol and ubiquinol in homogeneous solution and in aqueous dispersions of lipids: unrecognized consequences of lipid particle size as exemplified by oxidation of human low density lipoprotein. Proc. Nat. Acad. Sci. U.S.A. 90, 45–49.

Ioku, K., J. Terao, and N. Nakatani. 1992. Antioxidative activity of arbutin in a solution and liposomal suspension. Biosci. Biotech. Biochem. 56, 1658–1659.

Ishikawa, Y., K. Morimoto, and S. Iseki. 1991. Atrovenetin as a potent antioxidant compound from *Penicillium* species. J. Am. Oil Chem. Soc. 68, 666–668.

Jin, F., J. Leitich, and C. von Sonntag. 1993. The superoxide radical reacts with tyrosine-derived phenoxyl radicals by addition rather than by electron transfer. J. Chem. Soc. Perkin Trans. 2, 1583–1588.

Jitoe, A., T. Masuda, I. G. P. Tengah, D. N. Suprapta, I. W. Gara, and N. Nakatani. 1992. Antioxidant activity of tropical ginger extracts and analysis of the contained cucurminoids. J. Agric. Food Chem. 40, 1337–1340.

Jovanovic, S. V., S. Steenken, M. Tosic, B. Marjanovic, and M. G. Simic. 1994. Flavonoids as antioxidants. J. Am. Chem. Soc. 116, 4846–4851.

Kagan, J. 1966. The photochemical conversion of caffeic acid to esculetin. A model for the synthesis of coumarins *in vivo*. J. Am. Chem. Soc. 88, 2617–2618.

Kagan, V., E. Serbinova, and L. Packer. 1990. Antioxidant effects of ubiquinones in microsmes and mitochondria are mediated by tocopherol recycling. Biochem. Biophys. Res. Commun. 169, 851–857.

Kato, S., K. Shindo, H. Kawai, M. Matsuoka, and J. Mochizuki. 1993. Studies on free radical scavenging substances from microorganisms. III. Isolation and structural elucidation of a novel free radical scavenger, resorstatin. J. Antibiot. 46, 1024–1026.

Khanduja, K. L. and S. Majid. 1993. Ellagic acid inhibits DNA binding of benzo[a]pyrene activated by different modes. J. Clin. Biochem. Nutr. 15, 1–9.

Kikugawa. K., A. Kunugi, and T. Kurechi. 1990. Chemistry and implication of degradation of phenolic antioxidants. In B. J. F. Hudson, ed., *Food Antioxidants*. Elsevier, London. pp. 65–98.

Kitta, K., Y. Hagiwara, and T. Shibamoto. 1992. Antioxidative activity of an isoflavonoid, 2″-O-glycosylisovitexin, isolated from green barley leaves. J. Agric. Food Chem. 40, 1843–1845.

Kozlowska, H., M. Naczk, F. Shahidi, and R. Zadernowski, 1990. Phenolic acids and tannins in rapeseed and canola. In F. Shahidi, ed., *Canola Rapeseed: Production, Chemistry, Nutrition, and Processing Technology*. Van Nostrand Reinhold, New York, pp. 193–210.

Lambelet, P., J. Loliger, F. Saucy, and U. Bracco. 1992. Antioxidant properties of coenzyme-Q10 in food systems. J. Agric. Food Chem. 40, 581–584.

Landi, L., D. Florentini, L. Cabrini, C. Stefanelli, A. M. Secchi, and G. Pedulli. 1990. Are ubiquinones chain-breaking antioxidants? In G. Lenaz, O. Barnabei, A. Rabbi, and M. Battino, eds., *Highlights in Ubiquinone Research*. Tayor and Francis, London, pp. 262–265.

Larson, R. A. and A. L. Rockwell. 1980. Fluorescence spectra of water-soluble humic materials and some potential precursors. Arch. Hydrobiol. 89, 416–425.

Lu, H. and G.-T. Liu. 1992. Anti-oxidant activity of dibenzocyclooctene lignans isolated from Schisandraceae. Planta Med. 58, 311–313.

Lunder, T. L. 1992. Catechins of green tea: antioxidant activity. In M.-T. Huang, C.-T. Ho, and C. Y. Lee, eds., *Phenolic Compounds in Food and Their Effects on Health*. II. Antioxidants and cancer prevention, Am. Chem. Soc. Sympos. Ser. #507, Chapter 9, pp. 114–120.

Lyon, C. K. 1972. Sesame: present knowledge of composition and use. J. Am. Oil Chem. Soc. 49, 245–257.

Mahgoub, S. E. O. and B. J. F. Hudson. 1985. Inhibition of the pro-oxidant activity of copper by primary antioxidants in lard. Food Chem. 16, 97–101.

Malterud, K. E., T. L. Farbrot, A. E. Huse, and R. B. Sund. 1993. Antioxidant and radical scavenging effects of anthraquinones and anthrones. Pharmacology 47, Suppl. 1, 77–85.

Marinova, E. M. and N. V. Yanishlieva. 1992. Inhibited oxidation of lipids. II. Comparison of the antioxidative properties of some hydroxy derivatives of benzoic and cinnamic acids. Fat Sci. Technol. 94, 428–432.

Masaki, H., T. Atsumi, and H. Sakurai. 1994. Hamamelitannin as a new potent active oxygen scavenger. Phytochemistry 37, 337–343.

Matsuura, T. and H. Matsushima. 1970. Photoinduced reactions — XXXVI. Photosensitized oxygenation of 3-hydroxyflavones as a nonenzymatic model for quercetinase. Tetrahedron 26, 435–443.

Matsuzaki, T. and Y. Hara. 1985. Antioxidative activity of tea leaf catechins. Nippon Nog. Kag. Kaishi 59, 129–134.

Mukai, K., K. Kikuchi, and S. Urano. 1990. Stopped-flow kinetic study of the regeneration reaction of tocopheroxyl radical by reduced ubiquinone-10 in solution. Biochim. Biophys. Acta 1035, 77–82.

Mukai, K., H. Morimoto, S. Kikuchi, and S.-I. Nagaoka. 1993a. Kinetic study of free-radical scavenging action of biological hydroquinones (reduced forms of ubiquinone, vitamin K, and tocopherol quinone) in solution. Biochim. Biophys. Acta 1157, 313–317.

Mukai, K., H. Morimoto, Y. Okauchi, and S.-I. Nagaoka. 1993b. Kinetic study of reactions between tocopheroxyl radicals and fatty acids. Lipids 28, 753–756.

Mukai, K., S. Itoh, K. Daifuku, H. Morimoto, and K. Inoue. 1993c. Kinetic study of the quenching reaction of singlet oxygen by biological hydroquinones and related compounds. Biochim. Biophys. Acta 1183, 323–326.

Nakatani, N. and R. Inatani. 1984. Constituents of spices of the family Labiatae. Part III. Two antioxidative diterpenes from rosemary (*Rosmarinus officinalis* L.) and a revised structure for rosmanol. Agric. Biol. Chem. 48, 2081–2085.

Nakatani, N., R. Inatani, H. Ohta, and A. Nishioka. 1986. Chemical constituents of peppers (*Piper* spp.) and application to food preservation: naturally occurring antioxidative compounds. Environ. Health Perspect. 67, 135–142.

Nakayama, T., T. Kuno, M. Hiramitsu, T. Osawa, and S. Kawakishi. 1993. Antioxidative and prooxidative activity of caffeic acid toward H_2O_2-induced DNA strand breakage dependent on the state of the Fe ion in the medium. Biosci. Biotech. Biochem. 57, 174–176.

Namiki, M. 1990. Antioxidants/antimutagens in food. Crit. Rev. Food Sci. Nutr. 29, 273–300.

Neta, P. and S. Steenken. 1981. Phenoxyl radicals: formation, detection, and redox properties in aqueous solutions. In M. A. J. Rodgers and E. L. Powers, eds., *Oxygen and Oxy-radicals in Chemistry and Biology*. Academic Press, New York, pp. 83–88.

Neta, P., R. E. Huie, P. Maruthamuthu, and S. Steenken. 1989. Solvent effects in the reactions of peroxyl radicals with organic reductants: evidence for proton-transfer-mediated electron transfer. J. Phys. Chem. 93, 7654–7659.

Nick, J. A., C. T. Leung, and F. A. Loewus. 1986. Isolation and identification of erythroascorbic acid in *Saccharomyces cerevisiae* and *Lypomyces starkei*. Plant Sci. 46, 181–187.

Niki, E. 1991. Action of ascorbic acid as a scavenger of active and stable oxygen radicals. Am. J. Clin. Nutr. 54, 119S–124S.

Niki, E. and M. Matsuo. 1992. Rates and products of reaction of vitamin E with oxygen radicals. In M. G. Simic, K. A. Taylor, J. F. Ward, and C. von Sonntag, eds., *Vitamin E in Health and Disease*. Plenum, New York. pp. 235–243.

Nordström, C. G. 1968. Autoxidation of quercetin in aqueous solution. An elucidation of the autoxidation reaction. Suom. Kemist. B41, 351–353.

Okuda, T., T. Yoshida, and T. Hatano. 1992. Antioxidant effects of tannins and related polyphenols. In M.-T. Huang, C.-T. Ho, and C. Y. Lee, eds., *Phenolic Compounds in Food and Their Effects on Health.* II. Antioxidants and cancer prevention, Am. Chem. Soc. Sympos. Ser. #507, Chapter 7, pp. 87–97.

Osawa, T., S. Kumazawa, and S. Kawakishi. 1991. Prunusols A and B, novel antioxidative tocopherol derivatives isolated from the leaf wax of *Prunus grayana* Maxim. Agric. Biol. Chem. 55, 1727–1731.

Packer, L. and S. Landvik. 1989. Vitamin E: introduction to its biochemistry and health benefits. Ann. N. Y. Acad. Sci. 570, 1–6.

Papadopolous, G. and D. Boskou. 1991. Antioxidant effect of natural phenols on olive oil. J. Am. Oil Chem. Soc. 68, 669–671.

Pratt, D. E. and P. M. Birac. 1979. Source of antioxidant activity of soybeans and soy products. J. Food Sci. 44, 1720–1722.

Pratt, D. E. and B. J. F. Hudson. 1990. Natural oxidants not exploited commercially. In B. J. F. Hudson, ed., *Food Antioxidants.* Elsevier, London. pp. 171–191.

Ramarathnam, N., T. Osawa, M. Namiki, and S. Kawakishi. 1989. Chemical studies of novel rice antioxidants. II. Identification of isovitexin, a C-glycosyl flavonoid. J. Agric. Food Chem. 37, 316–319.

Record, I. R., I. E. Dresoti, and J. K. McInerney. 1995. The antioxidant activity of genistein *in vitro.* J. Nutr. Biochem. 6, 481–485.

Ricardo da Silva, J. M., N. Darmon, Y. Fernandez, and S. Mitjavila. 1991. Oxygen free radical scavenger capacity in aqueous models of different procyanidins from grape seeds. J. Agric. Food Chem. 39, 1549–1552.

Rice-Evans, C. A., N. J. Miller, P. G. Bolwell, P. M. Bramley, and J. B. Pridham. 1995. The relative antioxidative activities of plant-derived polyphenolic flavonoids. Free Rad. Res. 22, 375–383.

Robinson, T. 1991. *The Organic Constituents of Higher Plants.* 6th ed. Cordus Press, North Amherst, MA.

Rougée, M. and R. Bensasson. 1986. Détermination des constantes de vitesse de désactivation de l'oxygène singulet en presence de biomolécules. Comp. Rend. Acad. Sci. Paris 20, 1223–1226.

Sayer, J. M., H. Yagi, A. W. Wood, A. H. Conney, and D. M. Jerina. 1982. Extremely facile reaction between the ultimate carcinogen benzo[a]pyrene 7,8-diol-9,10-epoxide and ellagic acid. J. Am. Chem. Soc. 194, 5562–5564.

Schenck, G. O. 1960. Selektivitat und typische Reaktionsmechanismen in der Strahlenchemie. Z. Elektrochem. 64, 997–1011.

Schuler, P. 1990. Natural oxidants exploited commercially. In B. J. F. Hudson, ed., *Food Antioxidants.* Elsevier, London. pp. 99–170.

Schwartz, K. and W. Ternes. 1992. Antioxidative constituents of *Rosmarinus officinalis* and *Salvia officinalis.* I. Determination of phenolic diterpenes with antioxidative activity amongst tocochromanols using HPLC. Z. Lebensmitt. Untersuch. Forsch. 195, 95–98.

Scott, B. G., J. Butler, B. Halliwell, and O. I. Aruoma. 1993. Evaluation of the antioxidant actions of ferulic acid and catechins. Free Rad. Res. Commun. 19, 241–253.

Scurlock, R., M. Rougee, and R. V. Bensasson. 1990. Redox properties of phenols: their relationships to singlet oxygen quenching and to their inhibitory effects on benzo[a]pyrene-induced neoplasia. Free Rad. Res. Commun. 8, 251–258.

Simic, M. 1988. Mechanism of inhibition of free-radical processes in mutagenesis and carcinogenesis. Mutat. Res. 202, 377–386.

Stocker, R., V. W. Bowry, and B. Frei. 1991. Ubiquinol-10 protects human low density lipoprotein more efficiently against lipid peroxidation than does α-tocopherol. Proc. Nat. Acad. Sci. USA 88, 1646–1650.

Stone, A. T. 1987. Reductive dissolution rates of manganese (III/IV) oxides by substituted phenols. Environ. Sci. Technol. 21, 979–988.

Struski, D. G. J. and A. Kozubek. 1992. Cereal grain alk(en)yl resorcinols protect lipids against ferrous ions-induced peroxidation. Z. Naturforsch. 47C, 47–50.

Sugiyama, H., K. P. Fung, and T. W. Wu. 1993. Purpurogallin as an antioxidant protector of human erythrocytes against lysis by peroxyl radicals. Life Sci. 53, 39–43.

Taira, J., T. Ikemoto, T. Yoneya, A. Hagi, A. Murakami, and K. Makino. 1992. Essential oil phenylpropanoids: useful as ·OH scavengers? Free Radical Res. Commun. 16, 197–204.

Takagi, T. and T. Iida. 1980. Antioxidant for fats and oils from canary seeds: sterol and triterpene alcohol esters of caffeic acid. J. Am. Oil Chem. Soc. 57, 326–330.

Takahama, U. 1987. Oxidation products of kaempferol by superoxide anion radical. Plant Cell Physiol. 28, 953–957.

Takamura, K. and M. Ito. 1977. Effects of metal ions and flavonoids on the oxidation of ascorbic acid. Chem. Pharm. Bull. 25, 3218–3225.

Taylor, W. I. and A. R. Battersby. 1967. *Oxidative Coupling of Phenols.* Marcel Dekker, New York.

Terao, J., H. Karasawa, H. Arai, A. Nagao, T. Suzuki, and K. Takama. 1993. Peroxyl radical scavenging efficiency of caffeic acid and its related phenolic compounds in solution. Biosci. Biotech. Biochem. 57, 1204–1205.

Thompson, M., C. R. Williams, and G. E. P. Elliott. 1976. Stability of flavonoid complexes of copper(II) and flavonoid antioxidant activity. Anal. Chim. Acta 85, 375–381.

Tønnesen, H. H. and J. V. Greenhill. 1992. Studies on cucurmin and cucurminoids. XXII. Cucurmin as a reducing agent and as a radical scavenger. Int. J. Pharmaceut. 87, 79–87.

Tønnesen, H. H., G. Smistad, T. Ågren, and J. Karlsen. 1993. Studies on cucurmin and cucurminoids. XXIII. Effects of cucurmin on liposomal lipid peroxidation. Int. J. Pharmaceut. 90, 221–228.

Tournaire, C., S. Croux, M.-T. Maurette, I. Beck, M. Hocquaux, A. M. Braun, and E. Oliveros. 1993. Antioxidant activity of flavonoids: efficiency of singlet oxygen ($^1\Delta_g$) quenching. J. Photochem. Photobiol. B19, 205–215.

Tsujimoto, Y., H. Hashizume, and M. Yamazaki. 1993. Superoxide radical scavenging activity of phenolic compounds. Internat. J. Biochem. 25, 491–494.

Uri, N. 1961. Mechanism of antioxidation. In W. O. Lundberg, ed., Autoxidation and antioxidants. Vol. 1. Interscience, New York. pp. 133–169.

Wei, H. C., R. Bowen, Q. Y. Cai, S. Barnes, and Y. Wang. 1995. Antioxidant and antipromotional effects of the soybean isoflavone genistein. Proc. Soc. Exp. Biol. Med. 208, 124–130.

Wei, H. C., Q. Y. Cai, and R. O. Rahn. 1996. Inhibition of UV light- and Fenton reaction-induced oxidative DNA damage by the soybean isoflavone genistein. Carcinogenesis 17, 73–77.

Yagi, K. and N. Ohishi. 1979. Action of ferulic acid and its derivatives as antioxidants. J. Nutr. Sci. Vitaminol. 25, 127–130.

Yamamoto, Y., E. Komuro, and E. Niki. 1990. Antioxidant activity of ubiquinol in solution and phosphatidylcholine liposomes. J. Nutr. Sci. Vitaminol. 36, 505–511.

Yamauchi, R., K. Kato, S. Oida, J. Kanaeda, and Y. Ueno. 1992. Benzyl caffeate, an antioxidative compound isolated from propolis. Biosci. Biotech. Biochem. 56, 1321–1322.

Yang, X.-W., M. Hattori, T. Namba, D.-F. Chen, and G.-J. Xu. 1992. Anti-lipid peroxidative effect of an extract of the stems of *Kadsura heteroclita* and its major constituent, kadsurin, in mice. Chem. Pharm. Bull. 40, 406–409.

Yanishlieva, N. V. and E. M. Marino. 1995. Effects of antioxidants on the stability of triacylglycerols and methyl esters of fatty acids of sunflower oil. Food Chem. 54, 377–382.

Yoshino, K., Y. Hara, M. Sano, and I. Tomita. 1994. Antioxidative effects of black tea theaflavins and thearubigin on lipid peroxidation of rat liver homogenates induced by *tert*-butyl hydroperoxide. Biol. Pharm. Bull. 17, 146–149.

6 NITROGENOUS ANTIOXIDANTS

I. URIC ACID AND OTHER PURINES

Uric acid or 8-hypoxanthine (6-1) is the most-studied and apparently the most active antioxidant having a purine structure. The acid has a pKa of 3.9, with the proton at O-8 being the most acidic, assuring that it exists in most cells entirely in the form of the anion, urate. Its relatively high polarity and water solubility restricts it to the aqueous regions of cells.

(6-1)

Uric acid occurs in high concentrations in the excretory products of many animals, including birds and reptiles, and for many years was considered a waste product with no known biochemical function. Still, however, it is present in many organs and tissues at high concentrations (in human blood plasma at around 300 μM, for example). This is quite close to the solubility limit for the compound.

Uric acid has since been demonstrated to be an effective antioxidant in biological systems containing DNA and lipids (Ames et al., 1981; Cohen, 1984; Niki et al., 1986). The antioxidant effectiveness is pH-dependent, at least in lipid dispersions; at pH 3 it is ineffective, at pH 5 it is of intermediate effectiveness, and above pH 7 it is highly effective (Niki et al., 1986). This suggests that the urate anion is the only effective quenching and/or scavenging species. Uric acid has also been repeatedly demonstrated to be readily oxidized by many reagents, including those thought to be important in biological damage. Purines with apparently similar structures are much less readily oxidized. In addition, other purines are far less effective biological antioxidants.

The hydroxyl radical reacts with uric acid at a high rate (second-order constant ca. 7×10^9 l/mol sec: Simic and Jovanovic, 1989). This rate constant is about one or two orders of magnitude greater than those observed for most biologically important compounds. Other purines, however, are attacked at comparable rates (adenine = 3×10^9 l/mol sec, guanine = 1×10^{10} l/mol sec). Uric acid appears to be inert to attack by superoxide or hydrogen peroxide.

The potential reaction of uric acid with peroxyl radicals has been studied quantitatively only with the trichloromethylperoxyl radical, $Cl_3COO\cdot$ (Willson et al., 1985). With this unusually reactive species, the rate constant is quite large (ca. 5×10^8 l/mol sec). This value is comparable to the rate constant (5.8×10^8 l/mol sec) measured for the same free radical with the vitamin E analogue, Trolox (Alfassi et al., 1993) and suggests that uric acid may have comparable reactivity. Reactions with other peroxyl radicals, such as those typical of cellular damage, have been implied on the basis of the inhibition of DNA damage and lipid peroxidation by uric acid (Cohen et al., 1984; Niki et al., 1986). Because of solubility considerations, it would appear that uric acid, like ascorbate, scavenges peroxyl radicals (or other reactive free radicals) in the aqueous phase before they can enter into micellar or liposomal environments to initiate lipid peroxidation.

Ozone reacts with uric acid in blood plasma, leading to a decrease in its concentration (Cross et al., 1992). Preliminary studies suggest that uric acid may be able to prevent ozone-induced damage to lipids, but not to proteins (Meadows et al., 1986).

The reactivity of uric acid with singlet oxygen is a matter of some dispute. Originally it was suggested to protect cells by efficiently quenching $^1O_2$ (Ames et al., 1981); more recent measurements, however, cast doubt on this assumption. In a "pure" $^1O_2$-generating system in which the sensitizer was separated from the target by a narrow air path, uric acid was virtually unreactive (Dahl et al., 1988). The latter authors suggested that the apparent protection from "$^1O_2$-induced" damage in solution might have been due to reactions of uric acid with sensitizer excited states or other reactive species.

The initial site or sites of attack on uric acid by oxidizing free radicals is not yet known with certainty. The O8 oxygen and the nitrogens at N7 or N9 have been proposed (Equation 6-1) (Maples and Mason, 1988; Simic and Jovanovic, 1989).

The ultimate *in vitro* reaction products of uric acid include allantoin (6-2), a compound that has also been demonstrated to occur in the bloodstream of animals under oxidant stress (Halliwell and Gutteridge, 1989). This would support the hypothesis that uric acid is actually undergoing attack by oxidants *in vivo*. Cyanuric, parabanic, oxalic and glyoxylic acids have also been identified *in vitro* (Kaur and Halliwell, 1990).

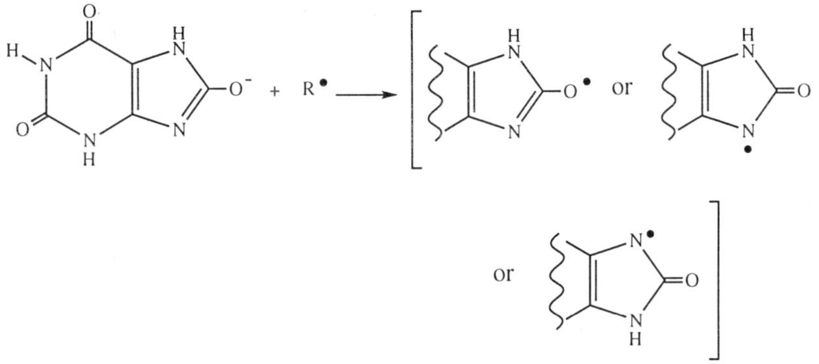

(Equation 6-1)

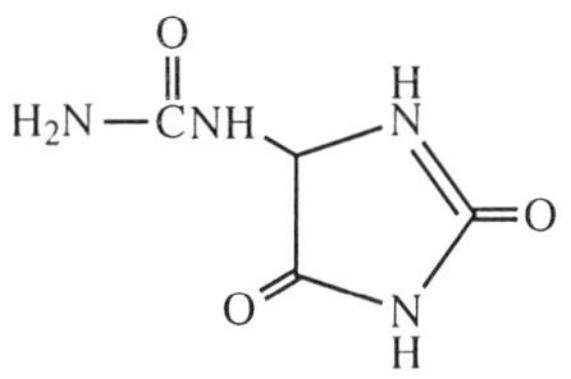

(6-2)

The urate radical may also contribute to antioxidant defense by its ability to "repair" other free radicals by intermolecular electron transfer:

$$\text{H-urate}^- + \text{R} \cdot \rightarrow \text{urate} \cdot + \text{RH}$$

(Equation 6-2)

The process has been demonstrated to occur with radicals derived from glutathione (Willson et al., 1985). This mechanism would be analogous to the well-studied ascorbate/vitamin E repair mechanism (Chapter 5). Uric acid may not be able to repair vitamin E; evidence is equivocal (Niki et al., *1986*). The urate radical, however, can be repaired by ascorbate (Simic and Jovanovic, 1989).

An additional mechanism of antioxidant activity that uric acid could display is its ability to act as a complexing agent for metal ions. It is able to inhibit the copper-promoted, Fenton-like decomposition of H_2O_2 to $HO\cdot$, for example (Halliwell, 1990).

II. AMINO ACID AND PEPTIDE DERIVATIVES

Amino acids have been variously reported to act as prooxidants, antioxidants, or to have little or no effect on oxidation reactions in various test systems. Among the wide variety of amino acid structures is a similarly wide range of charge characteristics at particular pHs, metal complexing capacity, redox potential, and tendency to interact with membranes having various charges and degrees of hydrophobicity. Accordingly, the wide range of responses noted by different authors using different experimental configurations is not too surprising.

Among the amino acids, one would expect that the best antioxidants would be found among those that were most readily oxidized (had the lowest redox potential). In general, this prediction is borne out. Most of the readily oxidized amino acids are those having heterocyclic and/or aromatic side chains such as tyrosine, tryptophan, and histidine. Cysteine, a thiol, is also readily oxidized to the thiyl radical.

The indole-containing amino acids and derivatives, tryptophan, melatonin, and tryptamine, are known to have antioxidative potential in some systems (Uemara et al., 1988; Christen et al., 1990; Poe-Geller et al., 1996). In appropriately substituted cases, electron transfer involving superoxide and subsequent cyclization to β-carboline structures may ensue. These compounds may also display good antioxidant activities (see Section 6-C).

In a study of the iron-induced lipid peroxidation of hepatic microsomes, several tryptamine derivatives were found to have good antioxidant efficiency (Table 6-1: Tse et al., 1991). Interestingly all of the compounds, regardless of substitution pattern, were about equally effective, and much more so than a hydroxylated derivative of tryptophan.

Probably the best-studied peptide antioxidant is glutathione (6-3). It, along with other sulfur-containing antioxidants, will be discussed in Chapter 7. The dipeptide, carnosine (β-alanyl-l-histidine) is one of the major lower-molecular weight constituents of muscle and brain tissue, often at levels that exceed 1 mM, and making up as much as 2% of the weight of skeletal muscle. Some N-methylated derivatives of carnosine, e.g., anserine and ofidine, are also present in these tissues. Carnosine has many physiological actions including wound-healing and antiinflammatory effects. The compound is also recognized to possess significant antioxidant activity (Boldyrev et al., 1987; Kohen et al.,

Table 6-1 Antioxidative Efficacies of Tryptamine Derivatives. IC_{50} = the Concentration (in μm) of the Substance that Inhibited 50% of Fe^{3+} ADP-Initiated Lipid Peroxidation in Rat Liver Microsomes

Compound	IC_{50}
Tryptamine	17.8
5-Hydroxytryptamine	5.6
5-Methoxytryptamine	10.1
Tryptophan	145

From Tse, S. Y., et al., 1991, *Biochem. Pharmacol.*, 42, 459–464.

1988) that is much greater than is provided by either of its amino acid moieties individually. Its activities in preventing oxidative stress, including lipid peroxidation, have been suggested to be due to metal ion chelation, to complexation with superoxide anion, or to reactions with lipid peroxidation intermediates. Although it is not particularly active toward peroxyl radicals, it has been demonstrated to inhibit several enzymes, including horseradish peroxidase and lysozyme, from free radical-mediated damage; the concentrations required were well within the range of observed intracellular levels (Salim-Hanna et al., 1991).

$$H_2N-\underset{COOH}{\underset{|}{CH}}-CH_2CH_2\overset{O}{\overset{\|}{C}}-NH-\underset{CH_2SH}{\underset{|}{CH}}\overset{O}{\overset{\|}{C}}NHCH_2COOH$$

(6-3)

III. ALKALOIDS AND RELATED COMPOUNDS

The group commonly known as alkaloids constitutes a tremendous variety of nitrogenous compounds; usually, but not always, of plant origin; usually, but not always, basic; and usually, but not always, heterocyclic. There have been few studies of alkaloids as antioxidant compounds, but under some conditions, some have been reported to be highly active.

An unprotonated, basic nitrogen atom might be expected to be a potent electron donor, and therefore a potentially good antioxidant, based on the rather low ionization and redox potentials of alkylamines. Amines are known to act as electron or hydride donors to a variety of free radicals including HO·, ROO·, and alkyl radicals (Buxton et al., 1988; Neta et al., 1990; Burton et al., 1996).

Many synthetic amine derivatives have been patented as oil stabilizers and corrosion inhibitors. However, most simple amines exist as fully protonated forms under ordinary physiological and environmental conditions. The pKa of methylamine, for example, is 10.7. (Heterocyclic or aromatic amines are usually much stronger acids, however; cf. pyridine, pKa = 5.2, and aniline, pKa = 4.6.) Protonation of most monobasic amines should almost completely suppress their ability to act as electron donors to another substrate. On the other hand, the positive charge on these compounds could aid in bringing them into conjunction with anionic regions of the cell such as DNA or phospholipid membranes. In fact DNA is normally strongly associated with organic cations, especially histones, basic proteins (high in lysine and arginine) that form near-stoichiometric, salt-like complexes with it in many cells. Histones can be displaced from DNA by other cations, including simple polyamines such as spermine (6-4).

$H_2N(CH_2)_3NH(CH_2)_4NH(CH_2)_3NH$

(6-4)

Polyamines such as spermine, spermidine (6-5), and putrescine (6-6) have been shown to accumulate in some plants exposed to elevated levels of UV-B (Kramer et al., 1991) and to exert protection in several stressed biological systems. Whether or not this is due to radical scavenging activity has not yet been resolved. Some authors attribute their protective activity against lipid peroxidation, for example, to their reactions with $\cdot O_2^-$ or $\cdot OH$ (Kitada et al., 1979; Drolet et al., 1986; Tadolini, 1988), whereas others have not found them to be efficient scavengers of free radicals (Bors et al., 1989), and others have proposed that as alkaline compounds they might quench acid-catalyzed peroxide decomposition (Løvaas, 1991). The cationic nature of these compounds has led to suggestions that coulombic interaction with anionic lipids or enzymes is itself a stabilizing mechanism (Schuber, 1989). However, in completely hydrophobic environments, they probably do exist as neutral compounds (Løvaas, 1991), and in fish oils and fish oil concentrates, spermine, spermidine, and putrescine were all shown to be much more powerful inhibitors of rancidity induction than gallate esters, BHA, or vitamin E. Rather high concentrations (ca. 1 mM) were usually necessary to observe significant increases in induction period, however.

$H_2N(CH_2)_3NH(CH_2)_4NH_2$

(6-5)

$H_2N(CH_2)_4NH_2$

(6-6)

A possible antioxidant role of alkaloids and related nitrogenous compounds could be as quenchers of singlet oxygen. As mentioned in Chapter 3, numerous amines are able to act as physical agents for deactivating $^1O_2$. Among naturally occurring compounds, several alkaloids of various structural types have been found to be potent inhibitors of $^1O_2$. Nicotine (6-7) was demonstrated to react with $^1O_2$ many years ago (Schenck and Gollnick, 1958), but there have been few subsequent investigations. Particularly effective, however, are indole alkaloids such as strychnine (6-8) and brucine (6-9) that incorporate a basic nitrogen atom in a rigid, cage-like structure (Gorman et al., 1984). These alkaloids have quenching rate constants that are greater than 10^8 l/mol sec, and are therefore comparably reactive to the most reactive amino acids. Other alkaloids tested were about an order of magnitude less reactive (Herlem and Ouannes, 1978; Larson and Marley, 1984). Most alkaloids, like simpler amines, appear to be strictly physical quenchers, and are not destroyed chemically by the process of quenching (Gorman et al., 1984). Therefore, in principle, they could inactivate many molecules of singlet oxygen per molecule of alkaloid.

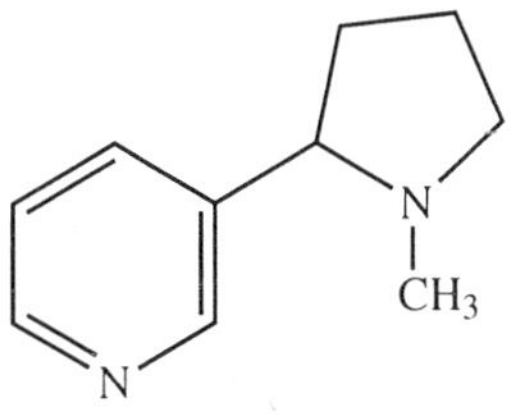

(6-7)

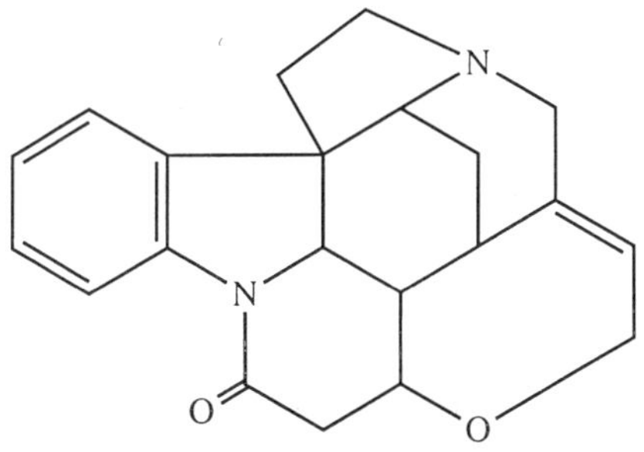

(6-8)

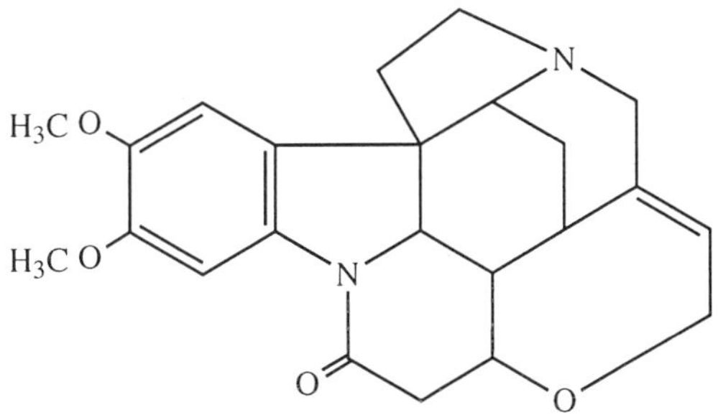

(6-9)

Boldine (6-10), an aporphine derivative, is the principal alkaloid found in the bark and leaves of the South American tree *Peumus boldo*. The mixed alkaloid fraction of the plant has found use in traditional medicines as well as being an ingredient of laxative and choleretic preparations. Boldine makes up about 0.1% of the leaves, but over 7% of the bark. The structure of boldine features a basic nitrogen atom, two aromatic methoxyl groups, and two phenolic hydroxyl functions.

(6-10)

Boldine was shown to inhibit autooxidation in several biological systems at concentrations of $10^{-4} - 10^{-5}$ M; for example, it prevented part of the peroxy radical-induced peroxidation of rat erythrocyte plasma membranes (Figure 6-1). Its activity appeared to be lower than that of the synthetic antioxidant, propyl gallate. The authors proposed that boldine scavenged alkylperoxy

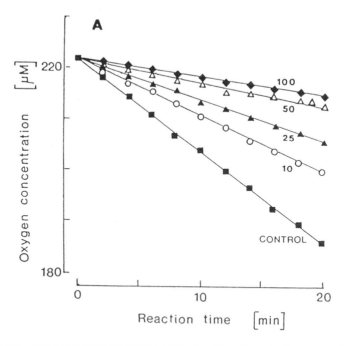

Figure 6-1 Effect of boldine on oxygen uptake by rat erythrocyte plasma membranes. Autooxidation initiated by an azo inhibitor. Boldine concentrations (μM) shown above each curve. (From Speisky, H., et al., 1991, Biochem. Pharmacol., 41, 1571–1581. Reprinted by permission from Elsevier Science, New York.)

radicals that would otherwise have initiated chain reactions (Speisky et al., 1991). It also protected against *t*-butyl hydroperoxide-induced glutathione oxidation and concomitant damage to rat hepatocytes at 2×10^{-4} *M* (Bannach et al., 1996). Boldine is also a quencher of singlet oxygen with a reported rate constant of about 2.4×10^7 l/mol sec (Larson and Marley, 1984). By analogy with substituted phenols (Chapter 5), the antioxidant activity of boldine is probably due to its ample supply of electron-donating (methoxyl, phenolic hydroxyl) substituents; however, few data are available from analogues of the compound to test this hypothesis.

Ubeda et al. (1992) and Seow et al. (1988) have also measured the antilipid peroxidative activity and anti-phagocytic activity of several benzylisoquinoline alkaloids including boldine. In the Ubeda study the most effective compound was apomorphine (6-11); they found it to have a greater activity than propyl gallate. Seow et al. measured the effects of the dimeric aporphine, tetrandrine (6-12) as an inhibitor of hydrogen peroxide production

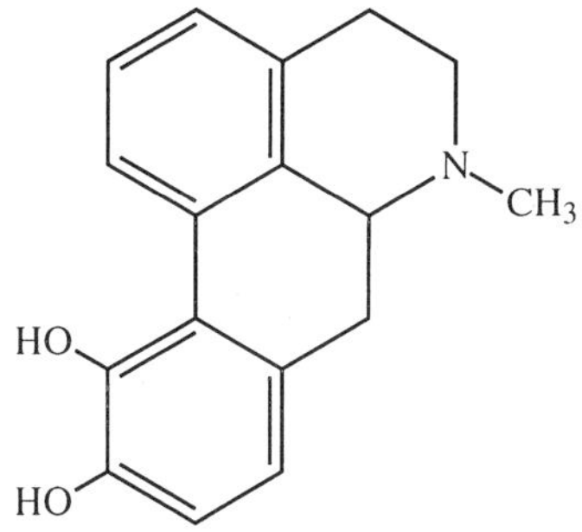

(6-11)

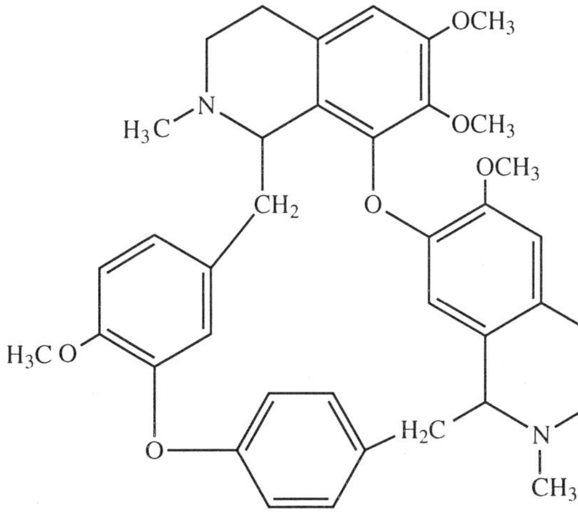

(6-12)

in neutrophils. Its activity was suggested to be as a scavenger of superoxide anions, although it is not clear by what mechanisms this could have occurred.

Another aporphine alkaloid, corytuberine (6-13), was also found to be a lipid antioxidant as well as a lipoxygenase inhibitor. Two somewhat related protoberberine alkaloids from the same plant (*Mahonia aquifolium*), namely oxyberberine (6-14) and columbamine (6-15), showed similar activities.

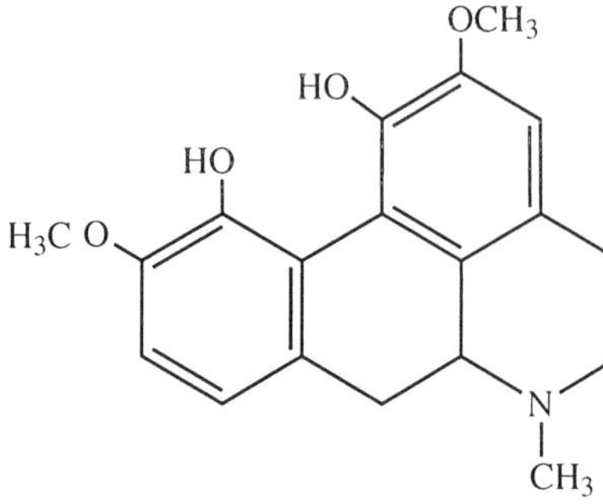

(6-13)

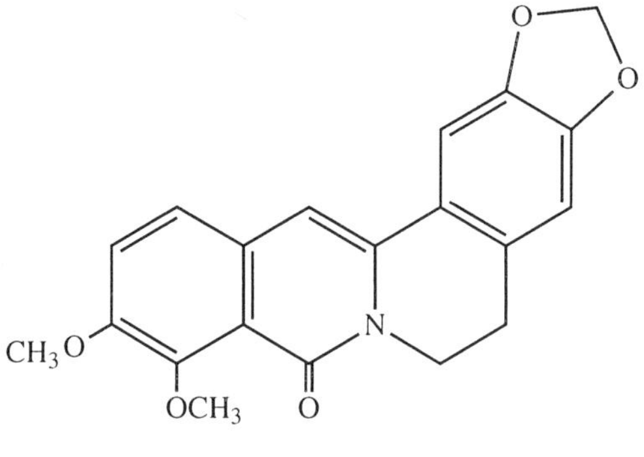

(6-14)

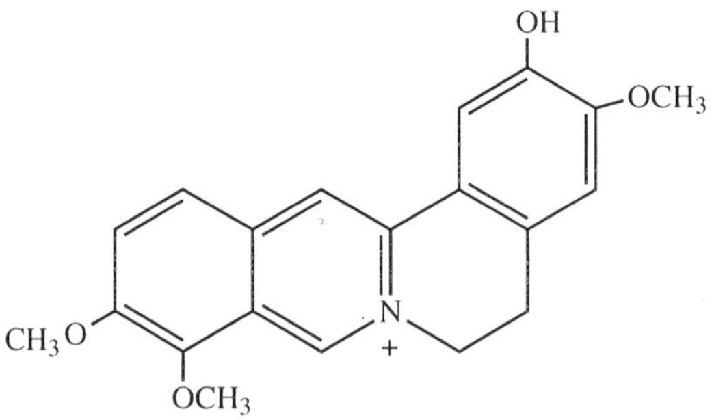

(6-15)

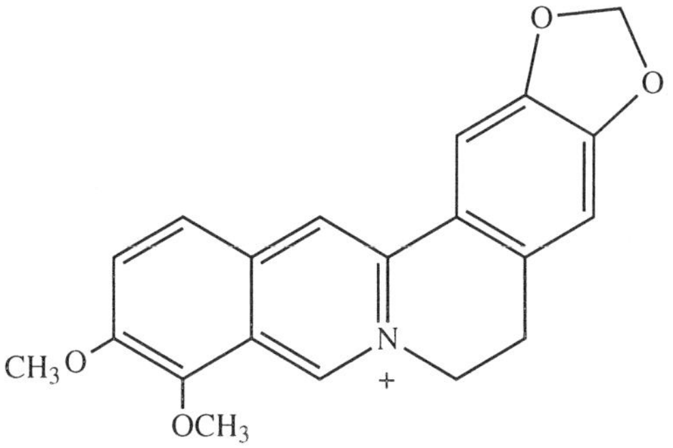

NITROGENOUS ANTIOXIDANTS 151

(6-16)

(Berberine itself [6-16] was inactive.) Because the peroxidation and enzyme inhibitory effects were well correlated, the authors proposed that the protective action of the compounds might have been due to their preventing lipid hydroperoxide accumulation (Misik et al., 1995).

Carazostatin (6-17) is a product from the microorganism *Streptomyces chromofuscus* with an indole alkaloid-like structure that shows potent antioxidant activity (Kato et al., 1989; Iwatsuki et al., 1993). The compound shows some intriguing similarities with vitamin E, including a sterically crowded phenolic hydroxy group across the ring from a heterocyclic substituent and a long hydrophobic chain. Although all the mechanistic details of the antioxidant activity of 6-17 are not known, the compound does undergo ready autooxidation in air, donates a hydrogen to the stable free radicals galvinoxyl and diphenylpicryl hydrazyl, and is readily oxidized electrochemically (anodic peak potential was ca. 74/V vs. the standard calomel electrode, a value about 56/V less than BHT: Jackson et al., 1991; Iwatsuki et al., 1993). A rational intermediate for the oxidation of 6-17 to products would be the iminoquinone 6-18. Derivatives of carazostatin that lacked the hydroxyl group were much less effective antioxidants (Iwatsuki et al., 1993). A simpler, synthetic carbazole, (6-19), an analog of carazostatin, is also a powerful inhibitor of lipid peroxidation in homogeneous solution and in phosphatidylcholine liposomes (Malvy et al., 1980; Figure 6-2).

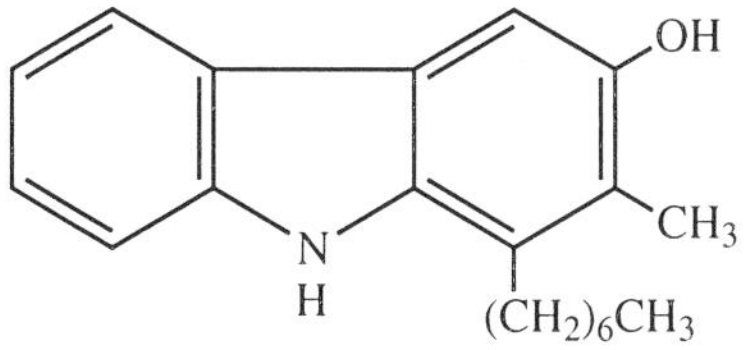

(6-17)

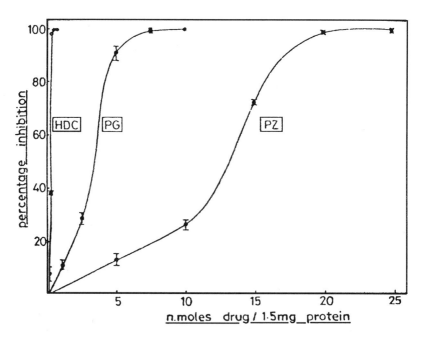

Figure 6-2 Inhibition of lipid peroxidation in rat liver microsomes by hydroxydimethylcarbazole (6-19; HDC) and other antioxidants. Peroxidation initiated by iron(II)-ascorbate. PG = propyl gallate, PZ = promethazine. (From Malvy, C., et al., 1980, Biochem. Biophys. Res. Commun., 95, 736. Reprinted by permission.)

NITROGENOUS ANTIOXIDANTS

Among the many actual plant indole alkaloids, only the β-carbolines have been tested so far (Tse et al., 1991). Both β-carboline (harmane 6-20, harmine 6-21, and harmol 6-22) and dihydro-β-carboline (harmaline 6-23 and harmalol 6-24) skeletal types were examined. Iron-induced lipid peroxidation in hepatic microsomes was shown to be inhibited to a greater or lesser degree by most of the alkaloid derivatives tested, though at widely varying efficiencies. Harmalol was demonstrated to have the greatest efficacy in this test, though it was far less effective than BHT. The concentrations of the alkaloids that inhibited 50% of the lipid peroxidation under the test conditions are listed in Table 6-2.

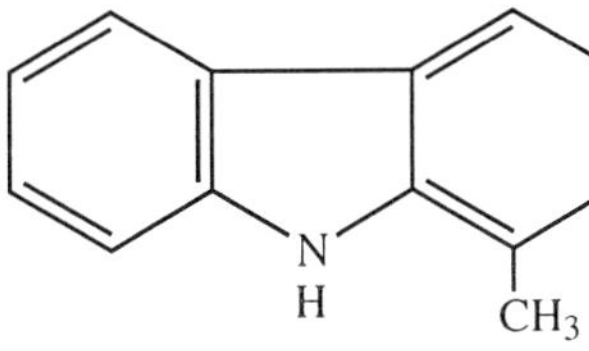

(6-20)

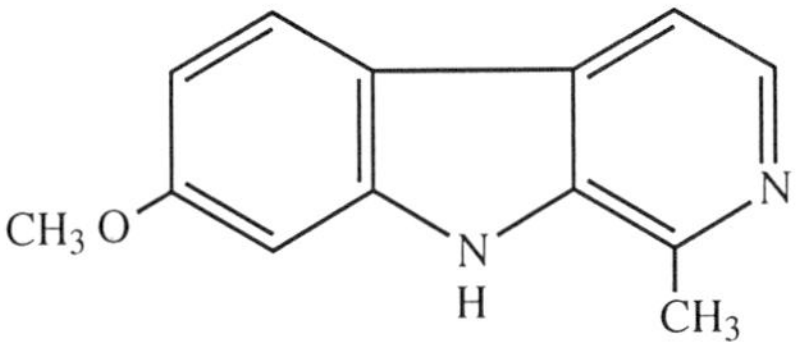

(6-21)

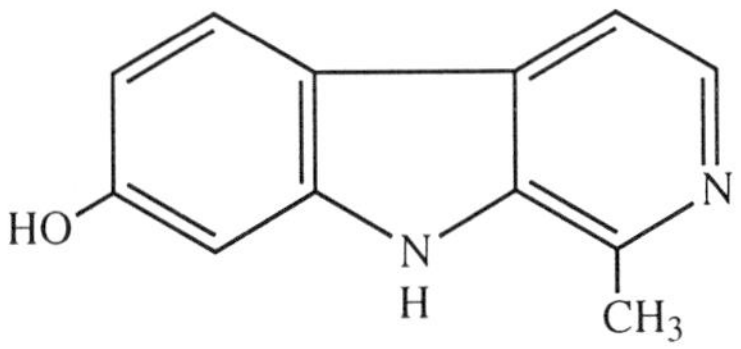

(6-22)

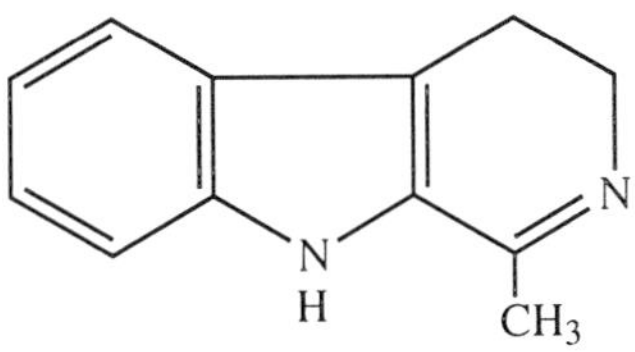

(6-23)

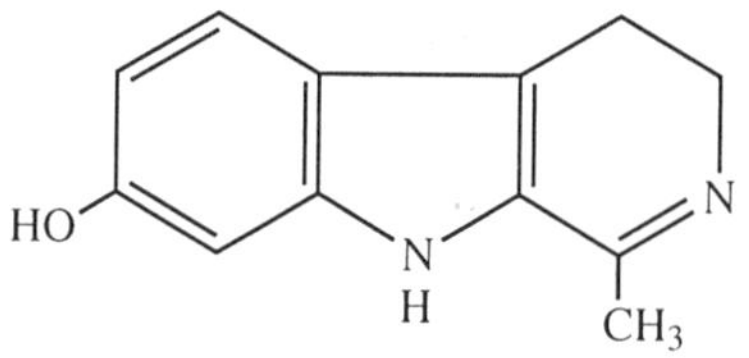

(6-24)

Table 6-2 Antioxidative Efficacies of Harmine Alkaloids. IC_{50} = the Concentration (in μm) of the Substance that Inhibited 50% of Fe^{3+} ADP-Initiated Lipid Peroxidation in Rat Liver Microsomes

Compound	IC_{50}
Harmalol	3.3
Harmane	11.7
Harmaline	10.4
Harmol	213
Harmine	>600

From Tse, S. Y., et al., 1991, *Biochem. Pharmacol.*, 42, 459–464.

It is evident that the dihydrocarbolines are superior to those having a fully aromatized quinoline ring, and that a structural substitution of hydroxyl for methoxyl increases antioxidant efficiency. The reasons for the differences in activity, as well as the relative importance of various potential oxidized radical intermediates (indolyl, phenoxyl) of these compounds and their resonance stabilization potential, are not entirely clear. Further and more quantitative studies should be undertaken with this series of compounds.

A dihydroindole derivative, stobadine (6-25), although not a naturally occurring compound (it is a synthetic, cardioprotective drug), nevertheless is a close relative of the β-carboline alkaloids. Its octanol-water partition coefficient of $10^{3.7}$ allows it to partition readily between lipid and aqueous phases (Kagan et al., 1993). It is a potent radical scavenger that donates an electron

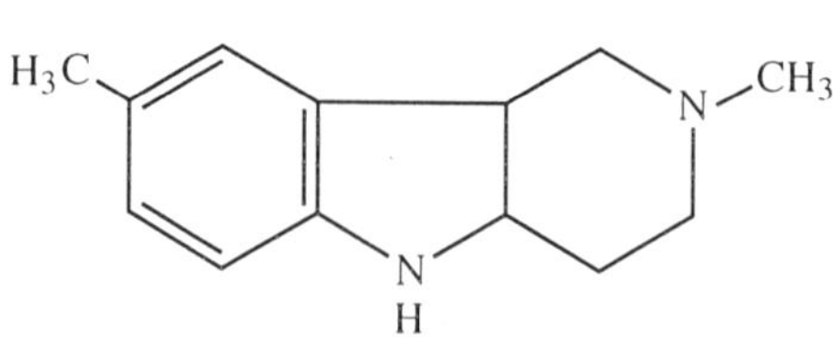

(6-25)

(from the indole nitrogen) to a variety of radicals including phenoxyl and peroxyl types, as well as HO· and Br_2^-· (Stefek and Benes, 1991; Steenken et al., 1992; Kagan et al., 1993). It reacts only quite slowly with superoxide, however (Kagan et al., 1993).

The redox potential of stobadine (0.58 V vs. NHE = 0.34 V vs. SCE) is considerably less than that of potent tertiary amine electron donors such as DABCO and is in the range of the tetraalkybenzidines, which are some of the most active organic electron donors (Kavarnos and Turro, 1986). Stobadine is also a quencher of singlet oxygen with a quenching rate constant of about 10^8 l/mol sec, which is comparable to the rate constants of the more reactive amino acids (Steenken et al., 1992).

Finally, three pyridine alkaloids have been examined for their ability to inhibit the peroxyl radical-catalyzed dimerization of tyrosine. Neither coniine (6-26), which has a pyridine ring that should be able to react with ROO· to produce an N-oxide (Mill et al., 1980), nor conyrine (6-27), a more basic compound, was an effective inhibitor of the reaction. However, nicotine (6-7), which incorporates both structural features, inhibited a significant fraction of the reaction. Preliminary kinetic calculations suggested that nicotine was about 5–10% as effective as ascorbate (Larson, 1995).

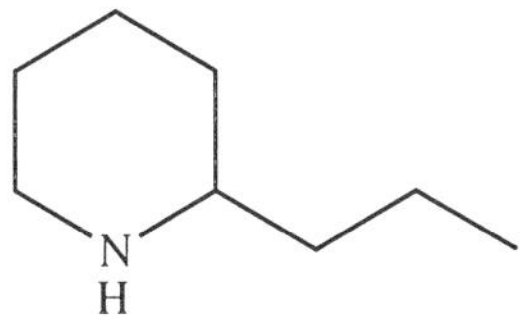

(6-26)

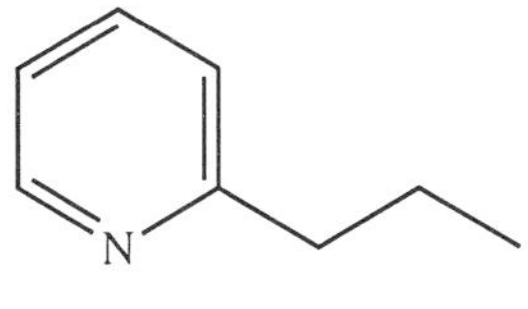

(6-27)

IV. TETRAPYRROLES

Bilirubin (6-28), a linear tetrapyrrolic bile pigment found in the blood, is bound to proteins such as albumin, and in this form it appears to be effective in inhibiting a number of free-radical-induced oxidation reactions, probably because of its reactivity with peroxyl radicals (Stocker et al., 1987). It is also a well-known sensitizer of singlet oxygen formation, but it is, in addition, a highly effective physical quencher. It also reacts with superoxide with a rate constant about one-tenth that of ascorbic acid, to afford bleached products. The mechanism of this reaction has not been well characterized (Halliwell and Gutteridge, 1989).

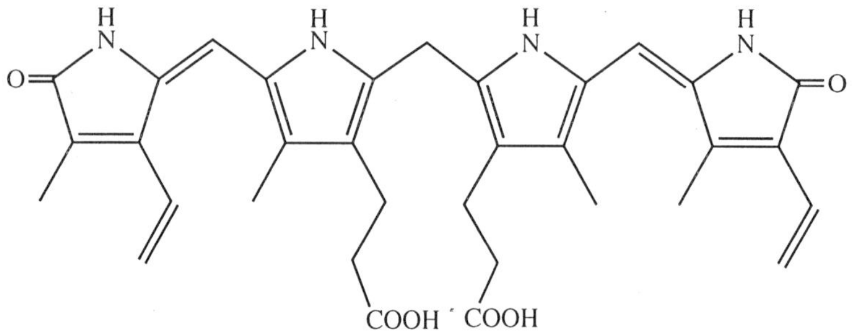

(6-28)

Tetrapyrroles may also be cyclized biosynthetically to form variously substituted derivatives of the cyclic ring system 6-29. The central group X may consist either of two hydrogen atoms or a metal cation. Natural tetrapyrroles are almost always iron or magnesium complexes (hemes and chlorophylls respectively). The substances are almost always colored, often with extremely strong absorption maxima in the visible region of the spectrum, that makes them invaluable as light-gathering compounds; the many forms of chlorophyll in the plant and bacterial kingdoms exemplify this role. The excited states of chlorophylls function in the plant as electron donors, converting light energy to chemical reducing power for the formation of organic carbon compounds from CO_2.

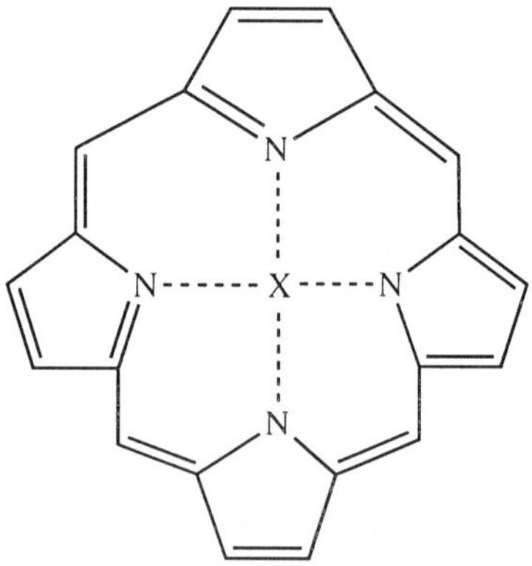

(6-29)

Photosynthesis aside, the bulk of the literature of free-radical chemistry and photochemistry of tetrapyrroles and their metal complexes (such as metalloporphyrins) addresses their pro-oxidant effects. Typically, the compounds are shown to be sources of hydroxyl radical, singlet oxygen, or superoxide upon illumination (Larson and Marley, 1994). In turn, they often decompose to metal-free or weakly colored derivatives. However, it is also true that chlorophyll and some related compounds, at least, appear to display some antioxidant activity.

In a test of fatty acid ester autooxidation, the target compounds were thermally incubated in the presence or absence of approximately 2×10^{-5} M chlorophyll A (6-30), chlorophyll B (6-31), pheophytin A (6-32), or pheophytin B (6-33). The reaction was monitored by measuring the formation of peroxides or carbonyl compounds in the sample. All of the compounds showed inhibitory effects (Figure 6-3: Endo et al., 1985a). Chlorophyll A showed the strongest antioxidant effects and was also effective, as was chlorophyll B, in delaying the peroxidation of soybean and rapeseed oil triglycerides. On the basis of the apperance (as seen by ESR) of a radical cation of chlorophyll in autooxidizing methyl linoleate solution, the authors concluded that it was acting as an electron or hydrogen atom donor to one or more free radicals forming during the reaction (Endo et al., 1985b). Chlorophyll was shown not to react with lipid hydroperoxides under the condition of the test.

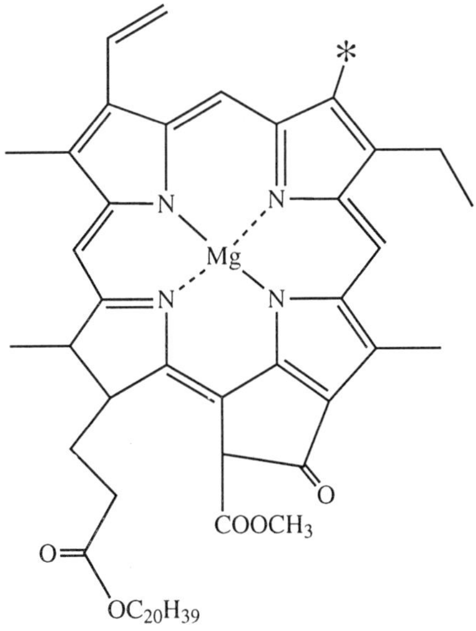

(6-30)

As in 6-30 but $-CHO$ at position $*$

(6-31)

As in 6-30 but without Mg (2 additional H).

(6-32)

As in 6-31 but without Mg (2 additional H).

(6-33)

A later study of chlorophyll A inhibition of lipid peroxidation showed a distinct solvent effect. In a polar solvent (heptanol), chlorophyll A showed a significant prolongation of the lag period in thermally autooxidizing methyl linoleate. Its activity was comparable to vitamin E or BHT. However, in the nonpolar solvent ethyl heptanoate, the chlorophyll was far less active than the two phenolic compounds. The authors concluded that a charge-transfer interaction between peroxy radicals (or anions) and chlorophyll (or its radical cation) was important in heptanol but not in the ester solvent (LeTutour et al., 1996).

Many marine organisms contain high levels of easily oxidized, unsaturated fatty acid derivatives. Accordingly, it has been suggested that they require efficient antioxidant protection; however, few compounds having traditional structural features associated with antioxidant activity have been identified from these organisms. A group from Shizuoka University (Sakata et al., 1990; Yamamoto et al., 1992; Watanabe et al., 1993), however, has identified a number of lipid-soluble constituents from several species of marine bivalves,

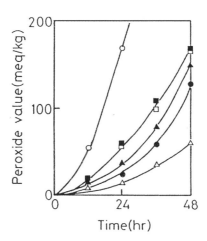

Figure 6-3 Inhibition of methyl linoleate thermal (30°C) autooxidation by chlorophyll derivatives. Open circles, control; solid squares, pheophytin B; open squares, pheophytin A; solid triangles, chlorophyll B; open triangles, chlorophyll A; solid circles, BHT. (From Endo, Y., et al., 1985a, J. Am. Oil Chem. Soc., 62, 1376. Reprinted by permission.)

including the clam, *Ruditapes phillipinarum*, which show unusually high activity. All of these compounds are chlorophyll relatives, including chlorophyllone-a (6-34), chlorophyllone lactone a (6-35), pyropheophorbide a (6-36), and methyl chlorophyllonate (6-37). These authors did not speculate on the mechanisms by which these compounds might act.

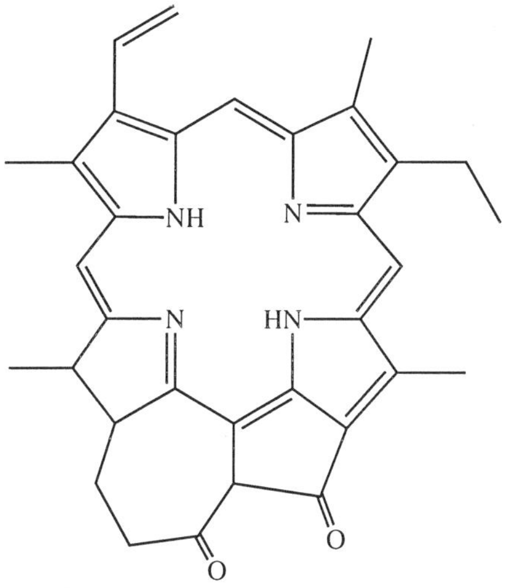

(6-34)

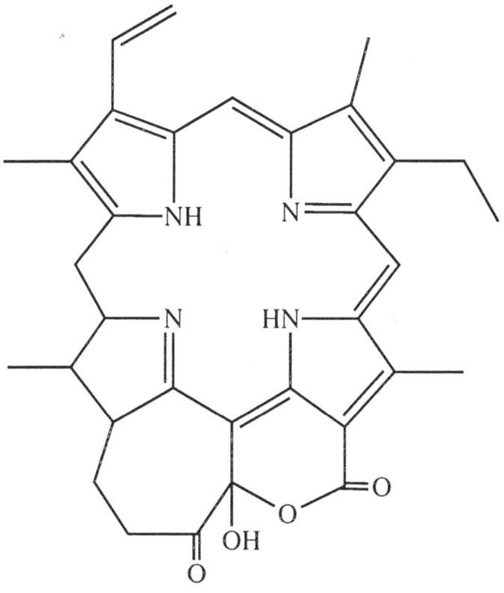

(6-35)

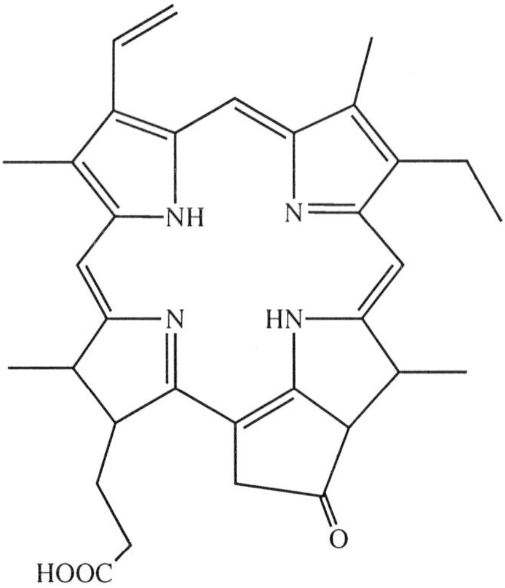

(6-36)

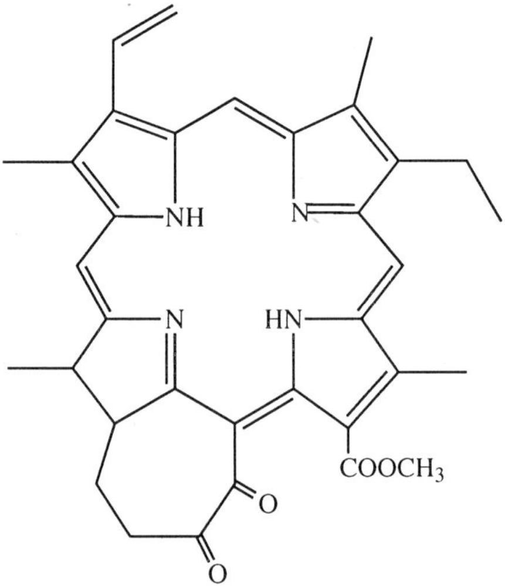

(6-37)

A very similar material, pyropheophytin a (6-38), has been identified as a constituent of the edible marine brown alga *Eisenia bicyclis*. Its antioxidant activity was demonstrated using standard techniques such as the ferric thiocyanate method (Figure 6-4; Cahyana et al., 1992).

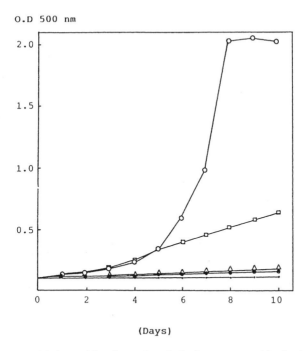

(6-38)

Figure 6-4 Antioxidative activity of pyropheophytin A as measured by the ferric thiocyanate method. Open circles, control; open squares, vitamin E; open triangles, pyropheophytin A. (From Cahyana, A., et al., 1992, Biosci. Biotech. Biochem., 56, 1535. Reprinted by permission.)

Interestingly, many of the chlorophyll degradation products identified in these marine organisms have also been reported to occur in heated or canned vegetables, where it is possible they could also have a role in preventing oxidative food degradation.

V. OTHER NITROGEN COMPOUNDS

For many years it has been inferred that antioxidants were formed when foods are heated, and that amino acids were implicated in the reactions that led to the active compounds. Although some investigators believed that polymeric materials were the principal antioxidants formed, others suggested that lower molecular weight products could also be important. Recent experiments in model systems have indicated some possible mechanisms for the process. For example, when a number of volatile, heterocyclic compounds derived from a thermal (Maillard) reaction between L-cysteine and D-glucose were tested for their ability to inhibit the autooxidation of aldehydes, it was found that methythiazoles and -oxazoles such as 6-39 and 6-40 possessed considerable activity (Eiserich et al., 1992). The authors suggested that the effectiveness of these heterocycles was correlated with the electron densities of their ring positions, which in turn would reflect the ability of oxidizing free radicals such as ·OH and ROO· to attack them.

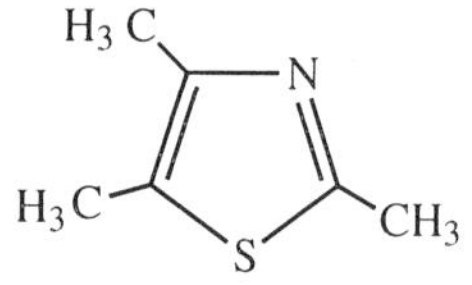

(6-39)

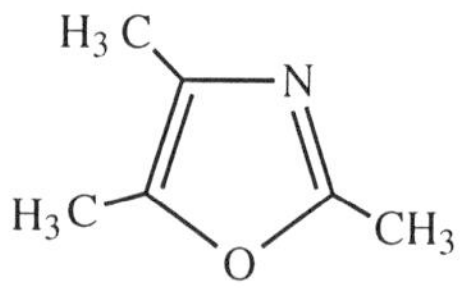

(6-40)

Another class of nitrogen heterocyclic compounds for which antioxidant activity has been claimed are pterin derivatives. Neopterin, and especially its 5,6,7,8-tetrahydro analogue 6-41, showed ability to prevent the copper(II)-induced oxidation of linoleic acid; at a concentration of 2×10^{-4} M, it was almost as effective as uric acid under the *in vitro* conditions employed. Other pterins were considerably less active (Kojima et al., 1992).

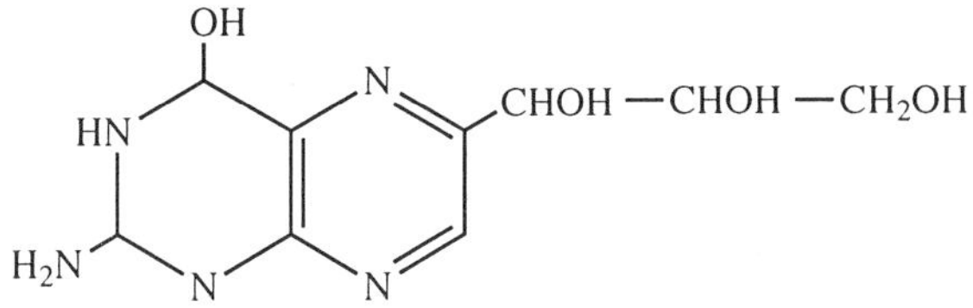

NITROGENOUS ANTIOXIDANTS

(6-41)

Shin-ya et al. (1991) report work on a compound with some structural similarity to the pterins, benthocyanin A, a phenazine derivative with a terpenoid side chain and a fused γ-lactone ring (6-42). The compound is produced by the mycelium of *Streptomyces prunicolor* and was reported, without other detail, to be "an order of magnitude more potent than vitamin E as a radical scavenger." Although it is not entirely clear why benthocyanin A should be so effective, it does have several intriguing structural elements such as a diphenylamine-like, potentially basic nitrogen atom that could act as an electron donor, and a quinone-like ring that might accept electrons. Further work on the mechanism of action of the compound is awaited.

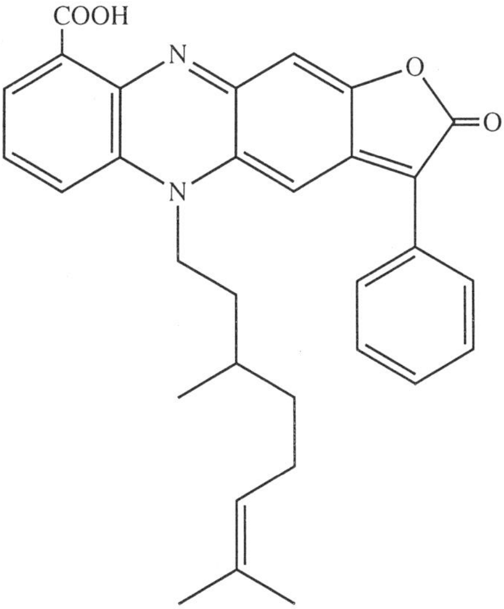

(6-42)

REFERENCES

Alfassi, Z. B., R. E. Huie, and P. Neta. 1993. Solvent effects on the rate constants for reaction of trichloromethylperoxyl radicals with organic reductants. J. Phys. Chem. 97, 7253–7257.

Ames, B. N., R. Cathcart, E. Schwiers, and P. Hochstein. 1981. Uric acid provides an antioxidant defense against oxidant- and radical-caused aging and cancer: a hypothesis. Proc. Natl. Acad. Sci. USA. 78, 6858–6862.

Bannach, R., A. Valenzuela, B. K. Cassels, L. J. Nunez-Vergara, and H. Speisky. 1996. Cytoprotective and antioxidant effects of boldine on *tert*-butyl hydroperoxide-induced damage to isolated hepatocytes. Cell Biol. Toxicol. 12, 89–100.

Boldyrev, A. A., A. M. Dupin, A. Y. Bunin, M. A. Babizhaev, and S. E. Severin. 1987. The antioxidative properties of carnosine, a natural histidine containing dipeptide. Biochem. Int. 15, 1105–1113.

Bors, W., C. Langebartels, C. Michel, and H. Sandermann. 1989. Polyamines as radical scavengers and protectants against ozone damage. Phytochemistry 28, 1589–1595.

Burton, A., K. U. Ingold, and J. C. Walton. 1996. Absolute rate constants for the reactions of primary alkyl radicals with aromatic amines. J. Org. Chem. 61, 3778–3882.

Buxton, G. V., C. L. Greenstock, W. P. Helman, and A. B. Ross. 1988. Critical review of rate constants for reactions of hydrated electrons, hydrogen atoms and hydroxyl radicals in aqueous solution. J. Phys. Chem. Ref. Data 17, 514–886.

Cahyana, A. H., Y. Shuto, and Y. Kinoshita. 1992. Pyropheophytin a as an antioxidative substance from the marine alga, arame (*Eisenia bicyclis*). Biosci. Biotech. Biochem. 56, 1533–1535.

Christen, S., E. Peterhans, and R. Stocker. 1990. Antioxidant activities of some tryptophan metabolites: possible implication for inflammatory diseases. Proc. Nat. Acad. Sci. U.S.A. 87, 2506–2510.

Cohen, A. M., R. E. Abendroth, and P. Hochstein. 1984. Inhibition of free radical-induced DNA damage by ascorbic acid. FEBS Lett. 174, 147.

Cross, C. E., P. A. Motchnik, B. A. Bruener, D. A. Jones, H. Kaur, B. N. Ames, and B. Halliwell. 1992. Oxidative damage to plasma constituents by ozone. FEBS Lett. 298, 269–272.

Dahl, T. A., W. R. Midden, and P. E. Hartman. 1988. Some prevalent biomolecules as sefenses against singlet oxygen damage. Photochem. Photobiol. 47, 357–362.

Drolet, G., E. B. Dumbroff, R. L. Legge, and J. E. Thompson. 1986. Radical scavenging properties of polyamines. Phytochemistry 25, 367–371.

Eiserich, J. P., C. Macku, and T. Shibamoto. 1992. Volatile antioxidants formed from an L-cysteine/D-glucose Maillard model system. J. Agric. Food Chem. 40, 1982–1988.

Endo, Y., R. Usuki, and T. Kaneda. 1985a. Antioxidant effects of chlorophyll and pheophytin on the autoxidation of oils in the dark. 1. Comparison of the inhibitory effects. J. Am. Oil Chem. Soc. 62, 1375–1378.

Endo, Y., R. Usuki, and T. Kaneda. 1985b. Antioxidant effects of chlorophyll and pheophytin on the autoxidation of oils in the dark. 2. The mechanism of antioxidative action of chlorophyll. J. Am. Oil Chem. Soc. 62, 1387–1390.

Gorman, A.A., I. Hamblett, K. Smith, and M.C. Standen. 1984. Strychnine: a fast physical quencher of singlet oxygen. Tetrahedron Lett., 25, 581–584.

Halliwell, B. 1990. How to characterize a biological antioxidant. Free Radical Res. Commun. 9, 1–32.

Halliwell, B. and J. M. C. Gutteridge. 1989. *Free Radicals in Biology and Medicine*. 2nd ed. Clarendon Press, Oxford, UK.

Herlem, D. and C. Ouannes. 1978. Oxydation photochimique d'une amine tertiare: la dregamine. Bull. Soc. Chim. France S1, 451–454.

Iwatsuki, M., E. Niki, and S. Kato. 1993. Antioxidant activities of natural and synthetic carbazoles. BioFactors 4, 1234–1280.

Jackson, P. M., C. J. Moody, and R. J. Mortimer. 1991. Synthesis and electrochemical properties of the naturally occurring free radical scavenger carazostatin. J. Chem. Soc. Chem. Commun. 2941–2944.

Kagan, V. E., M. Tsuchiya, E. Serbinova, L. Packer, and H. Sies. 1993. Interaction of the pyridoindole stobadine with peroxyl, superoxide, and chromanoxy radicals. Biochem. Pharmacol. 45, 393–400.

Kato, S., H. Kawai, T. Kawasaki, Y. Toda, T. Urata, and Y. Hayakawa. 1989. Studies on free radical scavenging substances from microorganisms. I. Carazostatin, a new free radical scavenger produced by *Streptomyces chromofuscus* DC118. J. Antibiot. 42, 1879–1881.

Kaur, H. and B. Halliwell. 1990. Action of biologically-relevant oxidizing species upon uric acid. Identification of uric acid oxidation products. Chem.-Biol. Interact. 73, 235–247.

Kavarnos, G. J. and N. J. Turro. 1986. Photosensitization by reversible electron transfer: theories, experimental evidence, and examples. Chem. Rev. 86, 401–449.

Kitada, M., K. Igarashi, S. Hirose, and H. Kitagawa. 1979. Inhibition by polyamines of lipid peroxide formation in rat liver microsomes. Biochem. Biophys. Res. Commun. 87, 388–394.

Kohen, R., Y. Yamamoto, K. C. Cundy, and B. N. Ames. 1988. Antioxidant activity of carnosine, homocarnosine, and anserine present in muscle and brain. Proc. Nat. Acad. Sci. U.S.A. 85, 3175–3179.

Kojima, S., T. Icho, Y. Kajiwara, and K. Kubota. 1992. Neopterin as an endogenous antioxidant. FEBS Lett. 304, 163–166.

Kramer, G. F., H. A. Norman, D. T. Krizek, and R. M. Mirecki. 1991. Influence of UV-B irradiation on polyamines, lipid peroxidation and membrane lipids in cucumber. Phytochemistry 30, 2101–2108.

Larson, R. A. 1995. Plant defenses against oxidative stress. Arch. Insect Biochem. Physiol. 29, 175–186.

Larson, R.A. and K. A. Marley. 1984. Quenching of singlet oxygen by alkaloids and related nitrogen heterocycles. Phytochemistry 23, 2351–2354.

Larson, R.A. and K. A. Marley. 1994. Oxidative mechanisms of phototoxicity. In J. O. Nriagu and M. S. Simmons, eds. *Environmental Oxidants*. Wiley, New York. pp. 269–317.

LeTutour, B., C. Brunel, and F. Quemeneur. 1996. Synergistic effect of chlorophyll A on the oxidation properties of vitamin E. New J. Chem. 20, 707–721.

Løvaas, E. 1991. Antioxidative effects of polyamines. J. Am. Oil Chem. Soc. 68, 353–358.

Malvy, C., C. Paoletti, J. F. A. Searle, and R. L. Willson. 1980. Lipid peroxidation in liver: hydroxydimethylcarbazole, a new potent inhibitor. Biochem. Biophys. Res. Commun. 95, 734–737.

Maples, K. R. and R. P. Mason. 1988. Free radical metabolite of uric acid. J. Biol. Chem. 263, 1709–1712.

Meadows, J., R. C. Smith, and J. Reeves. 1986. Uric acid protects membranes and linolenic acid from ozone-induced oxidation. Biochem. Biophys. Res. Commun. 137, 536–541.

Mill, T., D. G. Hendry, and H. Richardson. 1980. Free-radical oxidants in natural waters. Science 207, 886–887.

Misik, V., L. Bezakova, L. Malekova, and D. Kostalova. 1995. Lipoxygenase inhibition and antioxidant properties of protoberberine and aporphine alkaloids isolated from *Mahonia aquifolium*. Planta Med. 61, 372–373.

Neta, P., R. E. Huie, and A. B. Ross. 1990. Rate constants for reactions of peroxyl radicals in fluid solutions. J. Phys. Chem. Ref. Data 19, 413–513.

Niki, E., M. Saito, Y. Yoshikawa, Y. Yasmamoto, and Y. Kamiya. 1986. Oxidation of lipids. XII. Inhibition of oxidation of soybean phosphatidylcholine and methyl linoleate in aqueous dispersions by uric acid. Bull. Chem. Soc. Japan 59, 471–477.

Poe-Geller, B., R. J. Reiter, R. Hardeland, D. X. Tan, and L. R. Barlow-Walden. 1996. Melatonin and structurally related endogenous indoles act as potent electron donors and radical scavengers *in vitro*. Redox Rep. 2, 179–184.

Sakata, K., K. Yamamoto, H. Ishikawa, A. Yagi, H. Etoj, and K. Ina. 1990. Chlorophyllone-a, a new pheophorbide-related compound isolated from *Ruditapes phillipinarum* as an antioxidative compound. Tetrahedron Lett. 31, 1165–1168.

Salim-Hanna, M., E. Lissi, and L. A. Videla. 1991. Free radical scavenging activity of carnosine. Free Rad. Res. Commun. 14, 263–270.

Schenck, G. O. and K. Gollnick. 1958. Cinetique et inhibition de réactions photosensibilisées en presence d'oxygène moleculaire. J. Chim. Phys. 55, 892–898.

Schuber, F. 1989. Influence of polyamines on membrane functions. Biochem. J. 260, 1–10.

Seow, W. K., A. Ferrante, L. Si-Ying, and Y. H. Thong. 1988. Antiphagocytic and antioxidant properties of plant alkaloid tetrandrine. Int. Arch. Allergy Appl. Immunol. 85, 404–409.

Shin-ya, K., K. Furihata, Y. Hayakawa, H. Seto, Y. Kato, and J. Clardy. 1991. The structure of benthocyanin a, a new free radical scavenger of microbial origin. Tetrahedron Lett. 32, 943–946.

Simic, M. G. and S. V. Jovanovic. 1989. Antioxidation mechanisms of uric acid. J. Am. Chem. Soc. 111, 5778–5882.

Speisky, H., B. K. Cassels, E. A. Lissi, and L. A. Vedalia. 1991. Antioxidant properties of the alkaloid boldine in systems undergoing lipid peroxidation and enzyme inactivation. Biochem. Pharmacol. 41, 1571–1581.

Steenken, S., A. R. Sundquist, S. V. Jovanovic, R. Crockett, and H. Sies. 1992. Antioxidant activity of the pyridoindole stobadine. Pulse radiolytic characterization of one-electron-oxidized stobadine and quenching of singlet molecular oxygen. Chem. Res. Toxicol. 5, 355–360.

Stefek, M. and L. Benes. 1991. Pyridoindole stobadine is a potent scavenger of hydroxyl radicals. FEBS Lett. 294, 264–266.

Stocker, R., A. N. Glazer, and B. N. Ames. 1987. Antioxidant activity of albumin-bound bilirubin. Proc. Nat. Acad. Sci. U.S.A. 84, 5918–5922.

Tadolini, B. 1988. Polyamine inhibition of lipoperoxidation. Biochem. J. 249, 33–36.

Tse, S. Y. H., I.-T. Mak, and B. F. Dickens. 1991. Antioxidative properties of harmane and β-carboline alkaloids. Biochem. Pharmacol. 42, 459–464.

Ubeda, A., C. Montesinos, M. Paya, C. Terencio, and M. J. Alcaraz. 1992. Antioxidant action of benzylisoquinoline alkaloids. Free Rad. Res. Commun. 18, 167–175.

Uemara, T., M. Kanashiro, T. Yamano, K. Hirai, and N. Miyazaki. 1988. Isolation, structure, and properties of the β-carboline formed from 5-hydroxytryptamine by the superoxide anion-generating system. J. Neurochem. 51, 710–717.

Watanabe, N., K. Yamamoto, H. Ishikawa, A. Yagi, K. Sakata, L. S. Brinen, and J. Clardy. 1993. New chlorophyll-a related compounds isolated as antioxidants from marine bivalves. J. Nat. Prod. 56, 305–317.

Willson, R. L., C. A. Dunster, L. G. Forni, C. A. Gee, and K. J. Kittridge. 1985. Organic free radicals and proteins in biochemical injury: electron- or hydrogen-transfer reactions? Phil. Trans. Roy. Soc. Lond. B311, 545–563.

Yamamoto, K., K. Sakata, N. Watanabe, A. Yagi, L. S. Brinen, and J. Clardy. 1992. Chlorophyllonic acid A methyl ester, a new chlorophyll A related compound isolated as an antioxidant from short-necked clam, *Ruditapes phillipinarum*. Tetrahedron Lett. 33, 2587–2588.

7 SULFUR-CONTAINING ANTIOXIDANTS

I. GLUTATHIONE AND OTHER AMINO ACID DERIVATIVES

A. Glutathione

The cysteine-containing tripeptide glutathione (GSH, 7-1), one of the most important biological antioxidants, occurs in rather high concentrations (often 1–10 mM) in the cytosol of many types of cells. (On the other hand, some organisms and tissues contain little or none, apparently relying on other sorts of compounds for the same functions. Human blood plasma, for example, contains it only at about 20–30 µM.) In addition, it is localized in organelles such as chloroplasts. It is a key component of a variety of cellular mechanisms including detoxification of foreign metabolites, maintenance of growth rates in bacteria, and protection against gamma-radiation damage (Morse and Dahl, 1978; Briviba and Sies, 1994). Some microorganisms even excrete GSH into their surroundings, apparently using it as a defense against a variety of toxic materials (Owens and Hartman, 1985).

$$H_2N-CH-CH_2CH_2\overset{O}{\overset{\|}{C}}-NH-\overset{O}{\overset{\|}{CH\overset{}{C}NHCH_2COOH}}$$
$$\overset{|}{COOH} \qquad\qquad \overset{|}{CH_2SH}$$

(7-1)

Glutathione is poorly absorbed if ingested, and most animals synthesize it *in vivo*. The body maintains an active transport system for shuttling it to and from various organs.

The antioxidant biochemistry of glutathione has been summarized (Hausladen and Alscher, 1993). Like other antioxidants, it is readily oxidized; thiols such as GSH, in particular, react rapidly with many one-electron oxidants to form thiyl radicals:

$$GSH + R\cdot \rightarrow GS\cdot + RH$$

(Equation 7-1)

The thiyl radical is also formed enzymatically by horseradish peroxidase and similar catalysts, which use hydrogen peroxide to carry out a reaction similar to Equation 7-1:

$$GSH + H_2O_2 \rightarrow 2\ GS\cdot + H2O$$

(Equation 7-2)

The anion of glutathione is even more readily oxidized by one-electron transfer reactions to give the same free radical. Since the pKa of the thiol group in GSH is around 8.7, oxidation of the anion may become important in a number of environmental and biochemical milieux; for example, the pH of the chloroplast during photosynthesis is around 8.

Thiyl radicals are themselves quite reactive; for example, they may react with molecular oxygen to form potentially damaging species like $RSO_2\cdot$ or $RSO\cdot$ (Asmus, 1990). Under normal cellular conditions a self-termination reaction often occurs:

$$2\ GS\cdot \rightarrow GSSG$$

(Equation 7-3)

The dimer is known as "oxidized glutathione." The free radical can also couple with other thiols, such as cysteine residues in proteins. When this occurs, a radical anion is produced which may transfer an electron to molecular oxygen, forming superoxide:

$$GS\cdot + RS^- \rightarrow \cdot GSSR^- + O_2 \rightarrow GSSR + \cdot O_2^-$$

(Equation 7-4)

Within the cell the ratio of GSH to GSSG is kept high, often around 100/1. In abnormal conditions where GSH is depleted or GSSG builds up, the cell becomes very susceptible to membrane rupture, protein inactivation, and other phenomena characteristic of oxidative damage. It is presumed, for example, that one mechanism of quinone toxicity is due to a nucleophilic reaction (Figure 7-1), that not only depletes glutathione, but also generates a hydroquinone intermediate that is susceptible to rapid one-electron oxidation to produce superoxide (O'Brien, 1991). Toxicity caused by heavy metals, such as mercury, has also been proposed to be partly due to interactions between

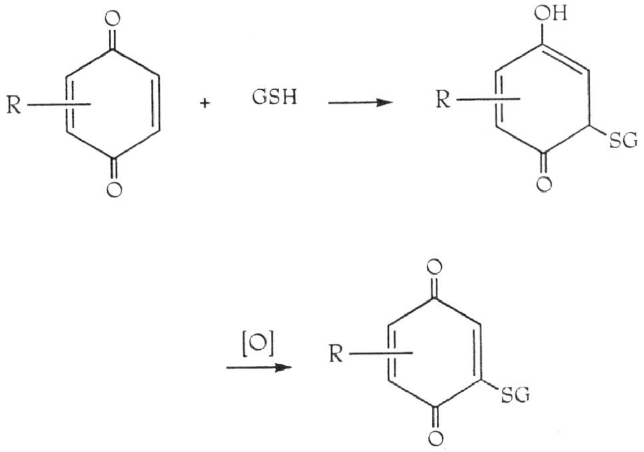

SULFUR-CONTAINING ANTIOXIDANTS

Figure 7-1 Nucleophilic reaction of quinones with thiols such as glutathione.

GSH and the metal ion; removal of GSH would make membranes such as that of the mitochondrion more susceptible to oxidative stress (Lund et al., 1991).

Another fate of the intermediate radical GS· is to be "repaired" by other intracellular H-donating antioxidants such as ascorbate or vitamin E. It has been suggested that one of the major functions of glutathione is to maintain ascorbate in the reduced state. The enzyme dehydroascorbate reductase, for example, makes use of GSH as a cofactor (Loewus, 1988). A related mechanism that may be important in membrane biochemistry is the ability of the water-soluble GSH to transfer an electron to membrane-associated vitamin E (Barclay, 1988).

The best-known function for glutathione is as a primary substrate for the selenium-containing enzyme, glutathione peroxidase, which catalyzes its reaction with hydrogen peroxide or lipid hydroperoxides:

$$ROOH + 2\ GSH \rightarrow ROH + GS\text{–}SG + H_2O$$

(Equation 7-5)

Oxidized glutathione is also formed in many of the nonenzymatic reactions of GSH, including its reactions with singlet oxygen, hydroxyl radicals, and superoxide. With $^1O_2$ it has a rate constant of 2.4×10^6 l/mol sec, about the same as quercetin and about one-sixth as great as vitamin E (Briviba and Sies, 1994). As with other water-soluble antioxidants, it is not clear how effective a defense mechanism it could be since $^1O_2$ has such a rapid rate of decay in water.

Similarly, with HO·, GSH reacts at rates close to diffusion control (rate constant ca. 10^9 l/mol sec, not too dissimilar from most biochemical substrates).

With superoxide, it reacts rather rapidly (k = 7.7 × 10^5 l/mol sec; Wefers and Sies, 1983) to afford a variety of redox products including hydrogen peroxide, GSSG, glutathione peroxides, and (perhaps) singlet oxygen.

Glutathione is also one of the few antioxidants that is capable of being directly oxidized by hydrogen peroxide (Florence, 1984) or hydroperoxides (Jocelyn and Dickson, 1980). GSSG is the product.

B. Carnosine

Another sulfur-containing peptide that has been suggested to possess antioxidant activity is carnosine (7-2). It is found in muscle tissue at levels from 1–60 mM. Physiologists have believed that carnosine's presence at such high levels could be related to its buffer activity; however, it also appears to be an effective antioxidant. Addition of carnosine to meat products, for example, leads to greatly improved storage stability due to inhibition of lipid oxidation (Decker and Crum, 1991). It is reported to have an unusual mechanism of action with autooxidizing phospholipids; while it does not greatly diminish overall oxygen uptake, it greatly decreases the formation of TBA-reactive carbonyl compounds (Kansci et al., 1994).

$$H_2N-CH_2CH_2-CONHCHCH_2-\text{(imidazole)}$$
$$|$$
$$COOH$$

(7-2)

Several investigators have shown that carnosine has relatively high activity with hydroxyl, peroxyl and other free radicals (Kohen et al., 1988; Aruoma et al., 1989; Pavlov et al., 1993; Babizhayev et al., 1994). Reactions of carnosine with superoxide are complex and pH-dependent, but seem to proceed via the preliminary formation of a charge-transfer complex between the two, which decomposes with the formation of disproportionation products of $\cdot O_2^-$. That is, carnosine appears to exhibit some superoxide dismutase-like activity (Pavlov et al., 1993). Other experiments indicate that it may be able to complex and inactivate transition metals such as copper and iron (Decker et al., 1992). Carnosine also reacts with lipid peroxides (Boldryev, 1986).

Carnosine, even at 1 mM, was not a particularly efficient quencher of singlet oxygen as measured in a competition experiment in which the rose bengal-sensitized disappearance of furfuryl alcohol was measured (Dahl et al., 1988).

C. Ergothioneine

The amino acid derivative l-ergothioneine (7-3) is a major sulfur-containing constituent of some fungi. It is not synthesized by mammals, but when ingested it is assimilated and its concentration maintained to the point where it can approach 1 mM concentrations in the liver, bone marrow, erythrocytes, and other tissues. (It is the principal thiol in human semen.) Evidence suggests that ergothioneine is likely to provide protection against several varieties of oxidative stress *in vivo*. For example, it inhibits lipid peroxide formation in mouse liver homogenates (Kawano et al., 1983).

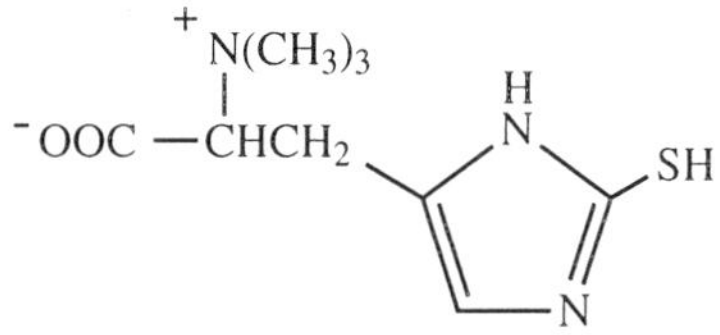

(7-3)

Ergothioneine undergoes one-electron oxidation by HO· and Cl$_3$COO·, and the oxidized radical intermediate is rapidly reduced, or repaired, by ascorbate. This interaction is similar to that observed between vitamin E radicals and ascorbate (Asmus et al., 1996). Hydroperoxides, but not H$_2$O$_2$, also react with ergothioneine (Arduini et al., 1992). It significantly reduces the mutagenicity of t-butyl hydroperoxide (and, to a lesser extent, cumene hydroperoxide) in *Salmonella* strain TA102 (Hartman and Hartman, 1987).

Ergothioneine showed good inhibitory activity toward rose bengal-photosensitized degradation of furfuryl alcohol in buffered aqueous solution, but was ineffective in a "pure singlet oxygen" exposure study in which the target compounds were physically separated from the sensitizer by air. The authors of the study concluded that it was interacting with a sensitizer excited state or some reactive species other than ^{1}O$_2$ (Dahl et al., 1988).

D. Ovothiol

The non-protein amino acid ovothiol A (7-4) is a thiol derivative of histidine that is found in marine animals and, oddly, also in parasitic protozoa (Turner et al., 1987; Steenkamp and Spies, 1994). The combination of readily oxidized functional groups in this compound make it an extremely effective antioxidant. In tests in which the compound was used as a scavenger for tyrosyl

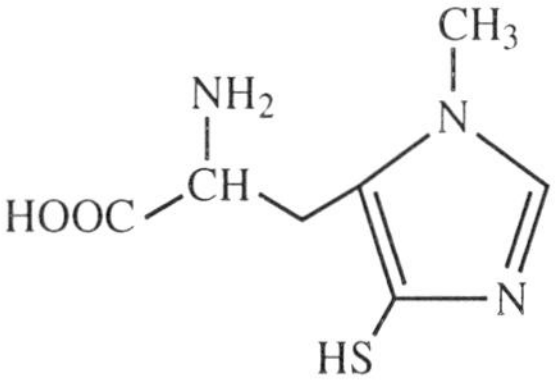

(7-4)

radicals in buffered aqueous solution, it was shown to have an efficiency superior to that of glutathione and nearly as high as those of the traditional antioxidants ascorbate, uric acid, and the water-soluble vitamin E analogue Trolox (Holler and Hopkins, 1989).

II. LIPOIC ACID AND RELATED SULFIDES AND POLYSULFIDES

Lipoic (or thioctic) acid (7-5) is biosynthesized from linoleic acid and occurs naturally in many organisms — microorganisms, plants, and animals — where it acts as an important coenzyme and growth factor. It sometimes occurs as the amide derivative. Therapeutically it is used in the treatment of liver diseases and as an antidote for *Amanita* mushroom poisoning. Recently it has been established to be a potent antioxidant *in vivo* and *in vitro*.

$$\text{structure: cyclic disulfide with } (CH_2)_4-COOH \text{ substituent}$$

(7-5)

Rats whose diets are supplemented with lipoic acid exhibit increased resistance to lipid peroxidation. However, 7-5 did not appear in this study to inhibit peroxy radical-induced lipid oxidation, suggesting that it has low reactivity toward ROO· in nonpolar phases (Kagan et al., 1990). In a later study of oxidative erythrocyte hemolysis, however, it inhibited the formation of adducts between the spin-trapping reagent DMPO and peroxyl (and alkoxyl) radicals derived from azo initiators (Constantinescu et al., 1994), possibly in the aqueous medium outside the cell. Conceivably, like some synthetic antioxidants, it also acts as an ROOH-destroying agent; for example, phenolic sulfides have been demonstrated to decompose *t*-butyl hydroperoxide (Jirackova et al., 1979).

Tests of the ability of the compound to inhibit autooxidation in the aqueous phase were also conducted (Suzuki et al., 1991). In these studies, lipoic acid prevented hydroxy-radical induced luminol chemiluminescence, and the compound's reduction product, dihydrolipoic acid (DHLA, 7-6) reacted with $\cdot O_2^-$ with an apparent second-order rate constant of 3.3×10^5 l/mol sec. DHLA is apparently able to transfer a second electron to $\cdot O_2^-$ to produce hydrogen peroxide. Another proposed action mechanism is the ability of lipoic acid to oxidize and deactivate iron(II), a necessary component of the Fenton reaction, and to be converted in turn to DHLA, also a good antioxidant. DHLA, unlike lipoic acid, was found to effectively deactivate peroxyl radicals both in the lipid phase and in the aqueous phase. It also donated an electron to ("repaired") ascorbate radicals, but not vitamin E phenoxyl radicals (Kagan et al., 1992).

$$\text{HS-CH}_2\text{-CH(SH)-(CH}_2)_4\text{-COOH}$$

(7-6)

Di- and polysulfides are relatively common naturally ocurring compounds in plants and even in some insect glands. Like other S compounds, they are readily oxidized, and have been shown to have antioxidant activity under some conditions. For example, an extract of garlic bulbs that protected membranes from lipid peroxidation included, as its most active constituents, several diallyl polysulfides $CH_2=CH-CH_2-S_n-CH_2-CH=CH_2$ (where n = 3–7). Whereas the trisulfide was relatively inactive, the polysulfides with n = 4–7 had roughly equal activity. At around $5 \times 10^{-5} M$, the heptasulfide inhibited 50% of the TBA-reactive substances formation in peroxidizing liver microsomes (Horie et al., 1992).

REFERENCES

Arduini, A., G. Mancinelli, G. L. Radatti, P. Hochstein, and E. Cadenas. 1992. Possible mechanism of inhibition of nitrite-induced oxidation of oxyhemoglobin by ergothioneine and uric acid. Arch. Biochem. Biophys. 294, 398–402.

Aruoma, O., M. J. Laughton, and B. Halliwell. 1989. Carnosine, homocarnosine, and anserine: could they act as antioxidants *in vivo*? Biochem. J. 264, 863–869.

Asmus, K- D. 1990. Sulfur-centered free radicals. Meth. Enzymol. 186, 168–180.

Asmus, K.-D., R. V. Bensasson, J.-L. Bernier, R. Houssin, and E. J. Land. 1996. One-electron oxidation of ergothioneine and analogues investigated by pulse radiolysis: redox reaction involving ergothioneine and vitamin C. Biochem. J. 315, 625–629.

Babizhayev, M. A., M. C. Seguin, J. Gueyne, R. P. Evstigneeva, E. A. Ageyeva, and G. A. Zheltukhina. 1994. L-Carnosine (*beta*-alanyl-L-histidine) and carcinine (*beta*-alanylhistamine) act as natural antioxidants with hydroxyl-radical-scavenging and lipid-peroxidase activities. Biochem. J. 304, 509–516.

Barclay, L. R. C. 1988. The cooperative antioxidant role of glutathione with a lipid-soluble and a water-soluble antioxidant during peroxidation of liposomes initiated in the aqueous phase and in the lipid phase. J. Biol. Chem. 263, 16138–16142.

Boldryev, A. A. 1986. On the biological role of histidine dipeptides. Biokhimiya 51, 1930–1943.

Briviba, K. and H. Sies. 1994. Nonenzymatic antioxidant defense systems. In B. Frei, ed., *Natural Antioxidants in Human Health and Disease*. Academic Press, San Diego. pp. 107–128.

Constantinescu, A., H. Tritschler, and L. Packer. 1994. *Alpha*-lipoic acid protects against hemolysis of human erythrocytes induced by peroxyl radicals. Biochem. Mol. Biol. Int. 33, 669–679.

Dahl, T. A., W. R. Midden, and P. E. Hartman. 1988. Some prevalent biomolecules as defenses against singlet oxygen damage. Photochem. Photobiol. 47, 357–362.

Decker, E. A. and A. Crum. 1991. Inhibition of oxidative rancidity in salted ground pork by carnosine. J. Food Sci. 56, 1179–1181.

Decker, E. A., A. D. Crum, and J. T. Calvert. 1992. Differences in the antioxidant mechanism of carnosine in the presence of copper and iron. J. Agric. Food Chem. 40, 756–759.

Florence, T. M. 1984. The production of hydroxyl radical from hydrogen peroxide. J. Inorg. Biochem. 22, 221–230.

Hartman, Z. and P. E. Hartman. 1987. Interception of some direct-acting mutagens by ergothioneine. Environ. Molec. Mutagen. 10, 3–15.

Hausladen, A. and R. G. Alscher. 1993. Glutathione. Chapter 1. In R. G. Alscher and J. L. Hess, eds., *Antioxidants in Higher Plants,* CRC Press, Boca Raton, FL. pp.1–30.

Holler, T. P. and P. B. Hopkins. 1989. A qualitative fluorescence-based assay for tyrosyl radical scavenging activity: ovothiol A is an efficient scavenger. Anal. Biochem. 180, 326–330.

Horie, T., S. Awazu, Y. Itakura, and T. Fuwa. 1992. Identified diallyl polysulfides from an aged garlic extract which protects the membranes from lipid peroxidation. Plant Med. 58, 468–469.

Jirackova, L., T. Jelinkova, J. Rotschova, and J. Pospisil. 1979. The mechanism of action of sulfur-containing antioxidants: reaction of phenolic sulfide with *t*-butyl hydroperoxide. Chem. Ind. (London) 11, 384.

Jocelyn, P. C. and J. Dickson. 1980. Glutathione and the mitochondrial reduction of hydroperoxides. Biochim. Biophys. Acta 590, 1–12.

Kagan, V., S. Khan, C. Swanson, A. Shvedova, E. Serbinova, and L. Packer. 1990. Antioxidant action of thioctic and dihydrolipoic acid. Free Rad. Biol. Med. 9S, 15.

Kagan, V. E., A. Shevdova, E. Serbovina, S. Khan, C. Swanson, R, Powell, and L. Packer. 1992. Dihydrolipoic acid, a universal antioxidant both in the membrane and in the aqueous phase. Reduction of peroxyl, ascorbyl, and chromanoxyl radicals. Biochem. Pharmacol. 44, 1637–1649.

Kansci, G., C. Genot, and G. Gandemer. 1994. Evaluation and antioxidant effect of carnosine on phospholipids by oxygen uptake and TBA test. Sci. Aliment. 14, 663–671.

Kawano, H., H. Murata, S. Iriguchi, T. Mayumi, and T. Hama. 1983. Studies on ergothioneine. XI. Inhibitory effect on lipid peroxide formation in mouse liver. Chem. Pharm. Bull. 31, 1682–1687.

Kohen, R., J. Jamamoto, K. C. Cundy, and B. N. Ames. 1988. Antioxidant activity of carnosine, homocarnosine, and anserine present in muscle and brain. Proc. Nat. Acad. Sci. U.S.A. 85, 3175–3179.

Loewus, F. A. 1988. Ascorbic acid and its metabolic products. In *The Biochemistry of Plants*, vol. 14, Academic Press, New York, pp. 85–107.

Lund, B.-O., D. M. Miller, and J. S. Woods. 1991. Mercury-induced H_2O_2 production and lipid peroxidation *in vitro* in rat kidney mitochondria. Biochem. Pharmacol. 42, S181–187.

Morse, M. L. and R. H. Dahl. 1978. Cellular glutathione is a key to the oxygen effect in radiation damage. Nature 271, 660–662.

O'Brien, P. J. 1991. Molecular mechanisms of quinone cytotoxicity. Chem.-Biol. Interact. 80, 1–41.

Owens, R. A. and P. E. Hartman. 1985. Export of glutathione (GSH) and other sulfur-containing compounds by *Salmonella typhimurium* and *Escherichia coli*. Environ. Mutagenesis 7 (Suppl. 3), 47.

Pavlov, A. R., A. A. Revina, A. M. Dupin, A. A. Boldryev, and A. I. Yaropolov. 1993. The mechanism of interaction of carnosine with superoxide radicals in water solution. Biochim. Biophys. Acta 1157, 304–312.

Steenkamp, D. J. and H. S. C. Spies. 1994. Identification of a major low-molecular-mass thiol of the trypanosomatid *Crithidia fasciculata* as ovothiol A: facile isolation and structural analysis of the bimane derivative. Eur. J. Biochem. 223, 43–50.

Suzuki, Y. J., M. Tsuchiya, and L. Packer. 1991. Thioctic and dihydrolipoic acid are novel antioxidants which interact with reactive oxygen species. Free Rad. Res. Commun. 15, 255–263.

Turner, E., R. Klevit, L. J. Hager, and B. M. Shapiro. 1987. Ovothiols, a family of redox-active mercaptohistidine compounds from marine invertebrate eggs. Biochemistry 26, 4028–4036.

Wefers, H. and H. Sies. 1983. Oxidation of glutathione by the superoxide radical to the disulfide and the sulfonate yielding singlet oxygen. Eur. J. Biochem. 137, 29–36.

8 CAROTENOIDS AND RELATED POLYENES

The carotenoids are yellow, orange, or red pigments of plant origin which are found in high concentrations in certain edible fruits or roots (mangoes, sweet potatoes, carrots, tomatoes, etc.). They are also found in the photosynthetic chloroplasts, where they function both as accessory light-gathering pigments (often bound to other lipids or proteins), and as antioxidants (Koyama, 1991). Their role as antioxidative compounds has been proved in numerous studies where lethal effects have been demonstrated in mutant organisms lacking them, or where carotenoid biosynthesis is inhibited by herbicides. In addition, carotenoids have been found in the eyes of some animals such as houseflies and some marine organisms, where their roles may include protection against light-induced damage.

Many hundreds of these compounds have been identified from natural sources. Structurally, they are long-chain polyisoprenes, usually with 40 carbon atoms in the chain. A small number are hydrocarbons, but the majority contain one or more oxygenated functional groups, and some occur in bound forms in association with carbohydrates or proteins. The compounds are unstable in storage, being especially susceptible to oxidizing agents, UV and visible light wavelengths, and alkaline reagents.

Carotenoids are required dietary constituents for animals, including humans, since some of them are precursors of the essential vitamin A. However, the occurrence of many carotenoids lacking vitamin A precursor activity in the body have directed the attention of researchers toward the other physiological roles these compounds may play. Over the past 20 years, carotenoids have received increasing attention for their purported activity as anticarcinogens (Krinsky, 1994; Khachik et al., 1995), although recently several human epidemiological studies have shown equivocal results, with some even suggesting that they may actually increase the risk of developing certain tumors such as lung cancers. There are stonger correlations between total diet (including total fruit and vegetable intake) and reduced cancer incidence than for carotenoid intake alone.

The antioxidant activities of carotenoids are due to their ease of reaction with a wide variety of oxidizing agents. Polyenes such as β-carotene (8-1), for example, react rapidly with free radicals under certain conditions. The derived radicals are rendered less energetic (more stable) by extensive resonance contributions from the long conjugated system of double bonds, making it less likely for the radical to take part in chain processes. β-Carotene also reacts rapidly with the triplet states of certain photosensitizing agents, especially ketonic sensitizers whose excited states are diradical-like. This process reduces the rate of chain initiation and is therefore an example of preventive antioxidation. In the chloroplast, this reaction probably also occurs, helping to quench undesirable excited states of chlorophylls that might otherwise produce singlet oxygen or other damaging species.

(8-1)

Many quenchers of excited states, such as β-carotene, are also potent quenchers of $^1O_2$. β-Carotene is the most reactive naturally occurring singlet oxygen quencher known (Foote et al., 1970), with a rate constant near that of diffusion control, exceeding $10^{10}/M$ sec. This rate constant exceeds that for the reaction of $^1O_2$ with most biologically important unsaturated fatty acids by 4–5 orders of magnitude, thus allowing a relatively low concentration of β-carotene to effectively protect membrane lipids from reactions of $^1O_2$ leading to peroxidation.

I. β-CAROTENE

The carotenes are a class of C40 terpenoid hydrocarbons found in almost all higher plants. The most abundant of these hydrocarbons, β-carotene, has a structure featuring two substituted cyclohexene (β-ionone) rings linked by a 22-carbon polyene chain. It is almost entirely insoluble in water, but readily soluble in hydrophobic environments and nonpolar solvents.

As a precursor of vitamin A (in fact, it is often referred to as provitamin A), β-carotene has found wide use in therapy. Oral administration of the compound has been prescribed in many instances of photosensitivity disorders, with varying degrees of success. Whereas the skins of patients taking large quantities of β-carotene become visibly yellow, with the compound becoming localized in the outer (epidermal) layers, it is a matter of dispute whether the compound exerts its action by physically absorbing potentially damaging sunlight radiation or whether its chemical antioxidant properties are more important (Sayre and Black, 1992).

β-Carotene has been reported over the past 30 years to exhibit high reactivity with electophiles and oxidants. The extensively conjugated double bond system readily donates an electron to oxidizing agents to afford highly stabilized radicals. Since there are many sites in the molecule that could take part in such reactions, the oxidation of β-carotene is extremely complex, and a great variety of products are formed. Only a few kinetic and mechanistic details of these reactions have been revealed.

Carotenoids, particularly β-carotene, scavenge free radicals under some conditions. Many studies have demonstrated that β-carotene inhibits autooxidation of lipids in biological tissues and in food products. Peroxyl radicals, in particular, have been shown either to add to the long chain of conjugated double bonds present in β-carotene and other carotenoids, or to take part in electron-transfer reactions giving rise to carbon-centered β-carotenyl free radicals:

$$\beta\text{-carotene} + ROO \cdot \rightarrow [ROO\text{-}\beta\text{-carotene} \cdot], \text{ or } [ROO^- + \beta\text{-carotene} \cdot]$$

(Equation 8-1)

For example, Packer et al. (1981) showed that the second-order rate constant for the reaction of β-carotene with $Cl_3COO \cdot$ was 1.5×10^9 l/mol sec, about a tenth of the diffusion-controlled rate. This rate constant is more than two orders of magnitude higher than those for unsaturated fatty acids (Willson, 1985). Others, however, have criticized this value as unrealistic for natural systems, due to the heightened reactivity of the trichloromethylperoxyl radical (Burton and Ingold, 1984). This suggestion is supported by more recent kinetic data for the reaction of β-carotene with initiator-derived peroxyl radicals, which give values in the 10^4–10^5 l/mol sec range (Ozhogina and Kasaikina, 1995). Although Burton and Ingold did not report a rate constant, they suggested that β-carotene was not a conventional chain-breaking antioxidant (that is, it did not compete effectively with the substrate for peroxyl radicals), nor was it a hydroperoxide-destroying (preventive) antioxidant. They suggested that its antioxidant activity was due to a preliminary H-atom transfer (or possibly an addition) to peroxyl radical (essentially, Equations 8-1–8-2),

$$ROO \cdot + \beta\text{-carotene} \rightarrow ROOH + \beta\text{-carotene} \cdot \text{ (or) } ROO\text{-}\beta\text{-carotene} \cdot$$

(Equation 8-2)

followed by a subsequent reaction between the β-carotenyl radical and another peroxyl radical;

ROO· + β-carotene· (or) ROO-β-carotene· → inactive products

(Equation 8-3)

The formation of ROO· addition products is supported by the isolation of products from azo-initiated radical reactions of β-carotene that incorporate one or two azo-derived groups (Liebler and McClure, 1996). Both types of reactions, electron donation and addition, were also observed in pulse radiolysis studies of $Cl_3COO·$-β-carotene reactions (Hill et al., 1995).

In the presence of higher oxygen concentrations, it was suggested that molecular oxygen, too, adds in a reversible manner to the carotenoid chain, producing peroxyl radicals with chain-carrying ability and limiting the ability of the compound to scavenge radicals efficiently. The authors concluded that β-carotene was an effective chain-breaking antioxidant only at oxygen concentrations of a tenth or so of normal atmospheric pressure (Burton and Ingold, 1984).

A different mechanistic suggestion was advanced by Kennedy and Liebler (1992), who argued that inactive autooxidation products of β-carotene accumulated in the medium, and were formed at faster rates at higher oxygen pressures. In any case, the possible pro-oxidant effect of β-carotene in oxygenated environments suggests a possible reason for its lack of protection, and even promoting effects, on cancers in the lung.

The reaction products of β-carotene with peroxyl radicals *in vitro* include epoxides, presumably formed by the addition mechanism of Equation 2-17 (El-Tinay and Chichester, 1970; Mordi et al., 1991; Kennedy and Liebler, 1991). Notably, this process should release alkoxyl radicals, which should be much more effective participants in chain reactions. Other products include a series of aldehydes and ketones that appear to arise from double-bond cleavage reactions at multiple sites along the chain (Handelman et al., 1991; Mordi et al., 1991). The mechanisms of the cleavage reactions are not clear at present, though cyclic endoperoxides are possible intermediates.

In vivo and *in vitro* studies of β-carotene protection against lipid and protein damage have been inconsistent. For example, it did not protect against lipoprotein oxidation initiated by water-soluble peroxyl radicals (Frei and Gaziano, 1993). Copper-induced autooxidation of low-density lipoproteins was inhibited by β-carotene in some studies (Jialal et al., 1991; Lavy et al., 1993), but in other studies it was found to be ineffective (Esterbauer et al., 1992; Frei and Gaziano, 1993). Another potential factor that may complicate *in vivo* investigations is the ability of β-carotene to very rapidly form addition products with thiyl radicals (RS·), such as that derived from glutathione. The rate constant for this reaction has been reported to be 2.2×10^8 l/mol sec (Everett et al., 1996).

The interactions of β-carotene with superoxide are complicated, and of uncertain importance for determining its antioxidant capacity. It has been shown that the radical anion of β-carotene is capable of transferring an electron to oxygen to form superoxide, but the reverse reaction does not occur. However, spectroscopic evidence suggests the formation of an unstable β-carotene-superoxide radical addition complex (Conn et al., 1992).

As mentioned above, β-carotene is an extremely good quencher of $^1O_2$, in fact probably the best of all naturally occurring substances tested so far. The quenching process appears to be mostly physical, with excess energy being dissipated either thermally to the medium or by reversible isomerization of one of the double bonds of the compound. In theory, therefore, a single β-carotene molecule should be able to inactivate many singlet oxygen molecules; one estimate suggests that the ratio is 1000:1 (Foote et al., 1970). The ability of carotenoids to quench $^1O_2$ falls off drastically as the number of conjugated couble bonds in the molecule decreases; those having more than 9 such bonds (β-carotene has 10) are ineffective quenchers (Foote and Denny, 1968).

II. OXYGEN-CONTAINING CAROTENOIDS

In addition to the carotenes, which are hydrocarbons, a great variety of oxygenated carotenoids (often referred to as xanthophylls) are found in plants, particularly in leaf chloroplasts. It has been suggested that in the intact plant, xanthophylls take up different positions in the membranes they are protecting than β-carotene does. The more polar characteristics of the alcoholic carotenoid derivatives, especially, would make it more likely that they would adopt an orientation that would line up with the fatty acid esters and tocopherols in the lipid bilayer, whereas the less polar carotene hydrocarbons might be localized well away from the membrane surface. Therefore, they might be able to inhibit different classes of oxidizing species.

Although β-carotene has been rather extensively studied, much less research, especially kinetically relevant research, has been performed on the antioxidant efficacy of oxidized carotenoids. It would seem that they would be structurally similar enough so that similar interactions with oxidants should take place, and the limited amount of information available suggests that this is the case. For example, lutein (8-2), like β-carotene, forms both electron-transfer and addition products with the peroxyl radical $Cl_3COO\cdot$(Hill et al., 1995). In some cases, alcoholic compounds such as lutein and zeaxanthin (8-3) have been shown to possess activity to inhibit lipid peroxidation and other biological oxidative processes (Terao, 1989; Lim et al., 1992; Khachik et al., 1995). From the shape of the kinetic curves, the compounds appear to be

(8-2)

(8-3)

acting as chain-breaking antioxidants. Xanthophylls have also been reported to have significant $^1O_2$ quenching activity, comparable to and perhaps even greater than β-carotene (DiMascio et al., 1989).

Carotenoid alcohols (such as zeaxanthin), and epoxy-alcohols (such as violaxanthin, 8-4) occur in plant tissues exposed to normal light levels. Studies of the responses of leaves to high light intensities have shown that illumination results in a loss of zeaxanthin, and increased violaxanthin concentrations. The reverse is seen in darkness, or in dark-adapted leaves. The phenomenon seems to be related to a protective process in which zeaxanthin specifically quenches certain excited states of chlorophyll. Excess photochemical excitation energy is therefore consumed in a non-destructive manner. Zeaxanthin appears to be produced from violaxanthin by an enzymatic de-epoxidation reaction. Therefore, although no direct evidence for the utilization of these compounds as traditional antioxidants in plants *in vivo* has been obtained, they apparently have important roles in protecting plants from light-induced stress (Young, 1991).

(8-4)

In several studies of radical scavenging, β-carotene was found to be less effective than some closely related carotenoids that bore carbonyl groups in the ionone rings. For example, astaxanthin (8-5) and canthaxanthin (8-6) were tested either in solution with autoxidizing methyl linoleate or a membrane model system with rat liver microsomal preparations. Lipid peroxidation, as observed either by malondialdehyde colorimetry or conjugated double bond formation, was suppressed more by the keto carotenoids than by β-carotene (Terao, 1989; Palozza and Krinsky, 1992; Palozza et al., 1996: cf. Figure 8-1). Similar results were observed for the same two compounds in a study of phospholipid peroxidation in liposomes, although in this case the order of inhibition was β-carotene > canthaxanthin > astaxanthin (Lim et al., 1992). Dietary canthaxanthin has also been suggested to function as an enhancer of vitamin E membrane concentration rather than as a direct-acting antioxidant (Mayne and Parker, 1989).

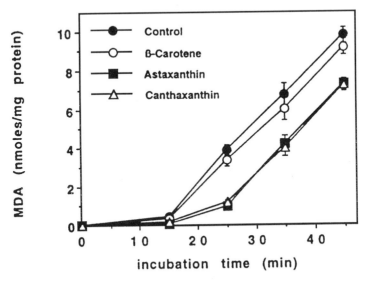

(8-5)

(8-6)

The greater efficiency of the keto carotenoids in lipid peroxidation has been attributed to a reduced activity of the stabilized polyene peroxyl radical toward propagation of the chain reaction (Terao, 1989).

Figure 8-1 Carotenoids as inhibitors of microsomal lipid peroxidation induced by an azo initiator. Antioxidants were present at a level of 10 nmol/mg protein. (From Palozza, P. and Krinsky, N. I., 1992, Arch. Biochem. Biophys., 297, 292. Reprinted by permission.)

III. RETINOL AND DERIVATIVES

Retinoids are naturally occurring and synthetic analogues of vitamin A. (The term "vitamin A" is often used in a physiological sense to refer to any of a group of compounds that have vitamin A activity, but will be used in this chapter to refer to retinol [8-7], or vitamin A alcohol.) The richest dietary sources of vitamin A itself are the liver oils of marine organisms such as sharks and halibut. The mammalian liver also contains enzymes that shorten the chains of dietary carotenoids and convert them to retinol; in some tissues, the vitamin is stored in esterified forms such as the acetate or palmitate. Deficiency of vitamin A is associated with numerous pathological conditions, especially involving the functions of the eye and the immune system. Oxidized forms of the vitamin also have physiological activity; for example, the aldehyde (retinal: 8-8) occurs within the eye as a protein-bound form, or opsin, and its isomerization is an important event in the physiology of vision. The acid (retinoic acid: 8-9) also occurs naturally, and as a synthetic form, under the trade name Retin-A, has been used in treatment of skin damaged by photoaging, acne, and other disorders.

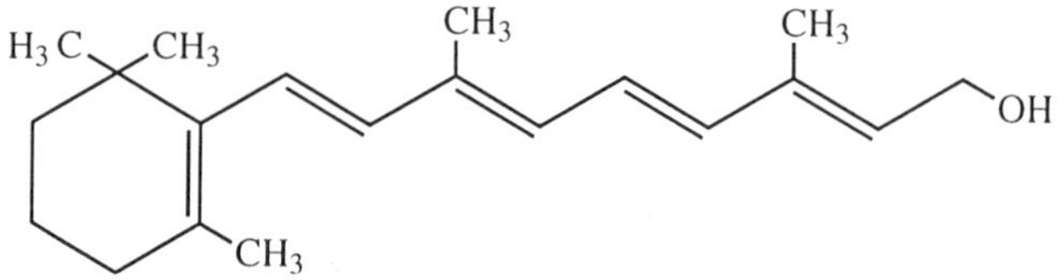

(8-7)

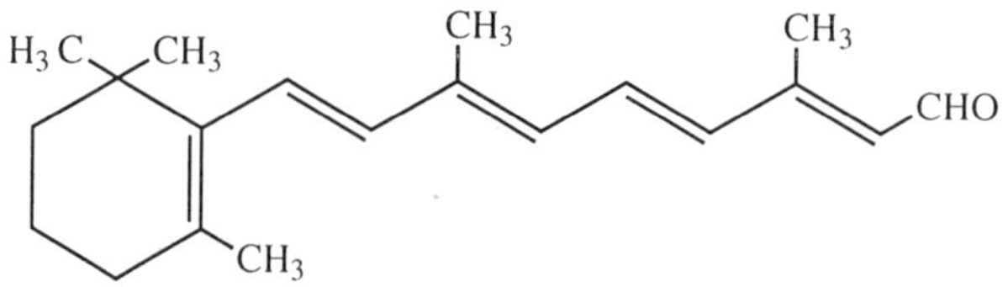

(8-8)

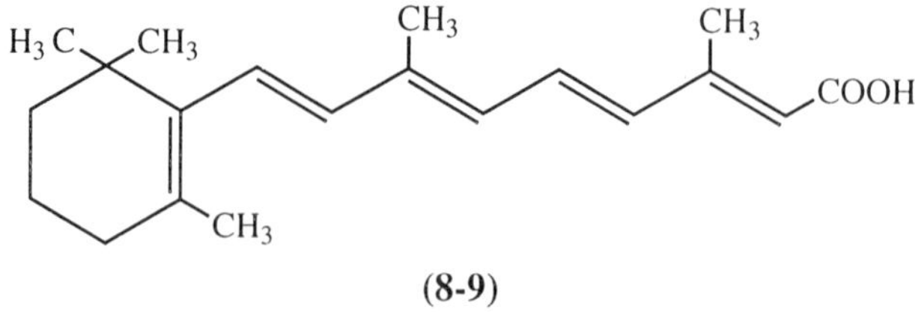

(8-9)

Retinol derivatives share many of the structural features of carotenoids, and the assumption has been that they could also exhibit antioxidant activities. Vitamin A is usually considered to be a dietary antioxidant, although it is

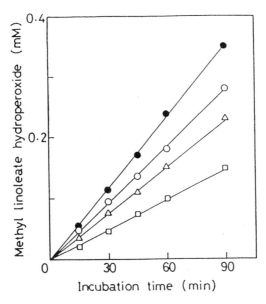

Figure 8-2 Effects of different concentrations of retinyl acetate of the formation of hydroperoxides from methyl linoleate (reaction induced by an azo initiator). Solid circles, control; open circles, 0.05 mM; open triangles, 0.1 mM; open squares, 1.0 mM. (From Yamauchi, R., et al., 1992, Biosci. Biotech. Biochem., 56, 1529. Reprinted by permission.)

almost certain that its effectiveness to protect against the usual oxidations that take place in the cell does not approach that of, for example, β-carotene. Because of the ability of these compounds to inhibit lipid peroxidation, at least to some extent, it is thought that they might react with peroxyl radicals *in vivo*. Evidence for such reactions was observed *in vivo* as inhibition of the rate of methyl linoleate peroxidation (induced by azo initiation) by retinyl acetate at $5 \times 10^{-5} - 1 \times 10^{-3}$ M concentrations (Figure 8-2). It is not clear, however, that the compounds could be present at high enough concentrations in the cell for these reactions to be significant. The products of the reaction were oxidized derivatives having epoxide, alcohol, and ether functionality in the polyene chain (Yamauchi et al., 1992). These compounds could have been formed by a mechanism involving addition of the peroxy radical to the polyene double bonds.

Vitamin A by itself is, as expected, usually a rather poor antioxidant (Gaby and Singh, 1991). It is stored intracellularly in protected environments, including specific sites in retinol-binding proteins (Olson, 1993). Therefore, it would appear that vitamin A is normally protected, rather than being protective. However, when it was incubated in combination with vitamin E in a liposomal oxidation system (initiated by azo compounds), the lag period for the mixture far exceeded the lag when either compound was used individually (Tesoriere et al., 1996). The authors were not sure of the mechanism of the observed synergism, but were able to rule out regeneration mechanisms in which one compound repaired the radical of the other.

REFERENCES

Burton, G. W. and K. U. Ingold. 1984. β-Carotene: an unusual type of lipid antioxidant. Science 224, 569–573.

Conn, P. F., C. Lambert, E. J. Land, W. Schalch, and T. G. Truscott. 1992. Carotene-oxygen radical interactions. Free Radical Res. Commun. 16, 401–408.

DiMascio, P., S. Kaiser, and H. Sies. 1989. Lycopene as the most efficient biological carotenoid singlet oxygen quencher. Arch. Biochem. Biophys. 274, 532–538.

El-Tinay, A. H. and C. O. Chichester. 1970. Oxidation of β-carotene: site of initial attack. J. Org. Chem. 35, 2290–2293.

Esterbauer, H., J. Gebicki, H. Puhl, and G. Jürgens. 1992. The role of lipid peroxidation and antioxidants in oxidative modification of LDL. Free Rad. Biol Med. 13, 341–390.

Everett, S. A., M. F. Dennis, K. B. Patel, S. Maddix, S. C. Kundu, and R. L. Willson. 1996. Scavenging of nitrogen dioxide, thiyl, and sulfonyl free radicals by the nutritional antioxidant β-carotene. J. Biol. Chem. 271, 3988–3994.

Foote, C. S. and R. W. Denny. 1968. Chemistry of singlet oxygen. VII. Quenching by β-carotene. J. Am. Chem. Soc. 90, 6233–6235.

Foote, C. S., R. W. Denny, L. Weaver, Y. Chang, and J. Peters. 1970. Quenching of singlet oxygen. Ann. N. Y. Acad. Sci. 171, 139–148.

Frei, B. and J. M. Gaziano. 1993. Content of antioxidants, preformed lipid hydroperoxides, and cholesterol as predictors of the susceptibility of human LDL to metal ion-dependent and -independent oxidation. J. Lipid Res. 34, 2135–2145.

Gaby, S. K. and V. N. Singh. 1991. β-Carotene. In *Vitamin Intake and Health*, S. K. Gaby, A. Bendich, V. N. Singh, and L. J. Machlin, eds., Marcel Dekker, NY. pp. 29–57.

Handelman, G. J., F. J. G. M. van Kuijk, A. Chatterjee, and N. I. Krinsky. 1991. Characterization of products formed during the autoxidation of β-carotene. Free Rad. Biol. Med. 10, 427–437.

Hill, T. J., E. J. Land, D. J. McGarvey, W. Schalck, J. H. Tinkler, and T. G. Truscott. 1995. Interactions between carotenoids and the $CCl_3OO\cdot$ radical. J. Am. Chem. Soc. 117, 8322–8326.

Jialal, I., E. P. Norkus, L. Cristol, and S. M. Grundy. 1991. β-Carotene inhibits the oxidative modification of low-density lipoprotein. Biochim. Biophys. Acta 1086, 134–138.

Kennedy, T. A. and D. C. Liebler. 1991. Peroxyl radical oxidation of β-carotene: formation of β-carotene epoxides. Chem. Res. Toxicol. 4, 290–295.

Kennedy, T. A. and D. C. Liebler. 1992. Peroxyl radical scavenging by β-carotene in lipid bilayers. Effect of oxygen partial pressure. J. Biol. Chem. 267, 4658–4663.

Khachik, F., G. R. Beecher, and J. C. Smith. 1995. Lutein, lycopene, and their oxidative metabolites in chemoprevention of cancer. J. Cell. Biochem. 22, 236–246.

Koyama, Y. 1991. Structures and functions of carotenoids in photosynthetic systems. J. Photochem. Photobiol. B9, 265–280.

Krinsky, N. I. 1994. Carotenoids and cancer: basic research studies. In B. Frei, ed., *Natural Antioxidants in Human Health and Disease.* Academic Press, San Diego, pp. 239–261.

Lavy, A., A. B. Amotz, and M. Aviram. 1993. Preferential inhibition of LDL oxidation by the all-*trans* isomer of β-carotene in comparison with 9-*cis* β-carotene. Eur. J. Clin. Chem. Clin. Biochem. 31, 83–90.

Liebler, D. C. and T. D. McClure. 1996. Antioxidant reactions of β-carotene: identification of carotenoid-radical adducts. Chem. Res. Toxicol. 9, 8–11.

Lim, B. P., A. Nagao, J. Terao, K. Tanaka, T. Suzuki, and K. Takama. 1992. Antioxidant activity of xanthophylls on peroxyl radical-mediated phospholipid peroxidation. Biochim. Biophys. Acta 1126, 178–184.

Mayne, S. T. and R. S. Parker. 1989. Antioxidant activity of dietary canthaxanthin. Nutr. Cancer 12, 225–236.

Mordi, R. C., J. C. Walton, G. W. Burton, L. Hughes, K. U. Ingold, and D. A. Lindsay. 1991. Exploratory study of β-carotene oxidation. Tetrahedron Lett. 32, 4203–4206.

Olson, J. A. 1993. Vitamin A and carotenoids as antioxidants in a physiological context. J. Nutr. Sci. Vitaminol. 39, S57–S65.

Ozhogina, O. A. and O. T. Kasakaina. 1995. β-Carotene as an interceptor of free radicals. Free Rad. Biol. Med. 19, 575–581.

Packer, J. E., J. S. Mahood, V. O. Mora-Arellano, T. F. Slater, R. L. Willson, and B. S. Wolfenden. 1981. Free radicals and singlet oxygen scavengers: reaction of a peroxy-radical with β-carotene, diphenyl furan, and 1,4-diazabicyclo[2.2.2]octane. Biochem. Biophys. Res. Commun. 98, 901–906.

Palozza, P. and N. I. Krinsky. 1992. Astaxanthin and canthaxanthin are potent oxidants in a membrane model. Arch. Biochem. Biophys. 297, 291–295.

Palozza, P., C. Luberto, P. Ricci, E. Sgarlata, G. Calviello, and G. M. Bartoli. 1996. Effect of β-carotene and canthaxanthin on *tert*-butyl hydroperoxide-induced lipid peroxidation in murine normal and tumor thymocytes. Arch. Biochem. Biophys. 325, 145–151.

Sayre, R. M. and H. S. Black. 1992. β-Carotene does not act as an optical filter in skin. J. Photochem. Photobiol. B12, 83–90.

Terao, J. 1989. Antioxidant activity of β-carotene-related carotenoids in solution. Lipids 24, 659–661.

Tesoriere, L., A. Bongiorno, A. M. Pintaudi, R. Dannar, D. Darpa, and M. A. Livrea. 1996. Synergistic interactions between vitamin A and vitamin E against lipid peroxidation in phosphatidylcholine liposomes. Arch. Biochem. Biophys. 326, 57–63.

Willson, R. L. 1985. Organic peroxy free radicals as ultimate agents in oxygen toxicity. In H. Sies, ed., *Oxidative Stress.* Academic Press, London, pp. 41–72.

Yamauchi, R., N. Mikaye, K. Kato, and Y. Ueno. 1992. Peroxyl-radical reaction of retinyl acetate in solution. Biosci. Biotech. Biochem. 56, 1529–1532.

Young, A. J. 1991. The photoprotective role of carotenoids in higher plants. Physiol. Plant. 83, 702–708.

Index

A

Active oxygen method, *see* Peroxide
  determination
Adenine
  oxidation products of, 10
  reaction with hydroxyl radical, 142
Aldehydes, *see* Carbonyl compounds
Alkaloids, 145–155
  acidity of, 145
  as singlet oxygen quenchers, 59, 146, 149
  β-carboline, 152–155
  reactions with alkyl radicals, 145
  reactions with hydroxyl radical, 145, 155
  reactions with peroxyl radicals, 145, 155
  reactions with superoxide, 150, 155
    see also Apomorphine, Boldine,
      Brucine, Columbamine, Coniine,
      Conyrine, Corytuberine, Harmane,
      Harmine, Harmalol, Harmol,
      Nicotine, Oxyberberine, Polyamines,
      Strychnine, Tetrandrine
Alkoxyl radicals (RO·), 11, 33, 60
  reactions of, 37–38
Alkyl radicals, 37, 67, 73, 76, 126
Amino acids
  complexation of metal ions by, 56–57, 74
  oxidation of, 8, 144
  reactions with singlet oxygen, 41
  reactions with superoxide, 43, 144
Anilines
  reaction with hydroxyl radical, 46
  reaction with peroxyl radicals, 71
Anthocyanins, 105
  see also Malvin, Nasunin
Anthrones, 124
Apigenin, 103
Apomorphine, 149
Arbutin, 122
Ascorbate (ascorbic acid), *see* Vitamin C
Astaxanthin, 184
Atrovenetin, 132
Autooxidation, 2
  initiated, kinetics, 67
Azo compounds, as radical initiators, 29,
  68–69, 112, 187

B

Benthocyanin A, 163
Benzyl caffeate, 116
BHA ("butylated hydroxyanisole"), 2, 98,
  125, 128, 146
BHT ("butylated hydroxytoluene"), 2, 4, 125,
  153
Bilirubin, 155–156
Boldine, 147–149
Browning reactions, *see* Maillard reactions
Brucine, 146
Butein, 108
tert-Butyl-4-methoxyphenol, *see* BHA
2,4-di-tert-Butyl-4-methylphenol, *see* BHT

C

Caffeic acid, 111–112
  esters, 116–117
  see also Chlorogenic acid
Canthaxanthin, 184
Capsaicin, 115
Carazostatin, 151
Carbohydrates, oxidation and oxidation
  products of, 11–13
Carbon black, 4
Carbonyl compounds (aldehydes and
  ketones), 37, 70, 121
  detection of, 18–19
Carboxylic acids
  as metal ion complexing agents, 57–58, 74,
    117
  see also Phenolic acids
Carnosic acid, 130
Carnosine, 144–145, 172
Carnosol, 130
Carotenoids, 39, 179–188
  as singlet oxygen quenchers, 59, 93, 180
  β-Carotene, 39, 42, 75, 93, 105, 180–183
    pro-oxidant effects, 182
    reactions with peroxyl radicals, 181–182
    reaction with singlet oxygen, 180, 183
    reactions with superoxide, 182
  Xanthophylls, 183–185

Carvacrol, 126
Catechin, 93, 105, 109
Catechins, 108–109
   reactions with hydroxyl radical, 109
   reactions with peroxyl radicals, 109
   reactions with singlet oxygen, 109
Catechols, 84
   see also Phenols
Cell membranes, 5
Cellulose, *see* Carbohydrates
Chain reactions, *see* Free radicals
Chalcones, 86, 108
Chelation therapy, 74
Chemical antioxidation, 52–53
Chlorogenic acid, 116
Chlorophylls, 157–159
Cholesterol, 5
Citric acid, *see* Carboxylic acids
Columbamine, 150
Complexation of metal ions, *see* Transition metal ions
Coniine, 155
Conjugated double bonds, in lipid peroxidation, 19
Conyrine, 155
Copper, *see* Transition metal ions
Corytuberine, 150
Coumarins, 112
Curcumin, 121

D

Daidzein, 107
Deferoxamine, 74
Diffusion, 78–79
Diperoxides, 13, 37, 39, 73
Diphenylpicrylhydrazyl, 28, 75–76
DNA, oxidation of, 9–11, 42
   see also Adenine, Guanine, Thymine, Uracil

E

EDTA (ethylenediamine tetraacetic acid), 55–56
Ellagic acid, 119
Endoperoxides, *see* Diperoxides
Epirosmanol, 131
Epoxides, 33, 70
Ergothioneine, 172–173

Erythorbic acid, 99
Ethylenediamine tetraacetic acid, *see* EDTA

F

Fenton reaction, 10, 33, 45, 143
Ferulic acid, 113–114
   esters, 116
   reaction with peroxyl radicals, 114
Fisetin, 93, 105
Flavonoids, 84, 100–106
   reactions with peroxyl radicals, 90–92
   reactions with singlet oxygen, 93–94
   reactions with superoxide, 88–89, 105
   see also Apigenin, Catechin, Fisetin, Hesperetin, Isovitexin, Kaempferol, Luteolin, Myricetin, Naringenin, Phenols, Quercetin, Robinetin, Rutin
Folic acid, 13
Free radicals, 25–38
   chain reactions, 29–38
      kinetics, 68–73, 76–78
      lag period method, 68, 78
   dimerization of, 31, 34, 71–72
   disproportionation, 34, 84
   formation of, 25–31
   reactions of, 31
   see also Alkyl, Alkoxyl, Hydroxyl, Peroxyl, Phenoxyl, Thiyl radicals; Superoxide
French paradox, 115
Furans, 12, 40, 59

G

Gallic acid, 114–115
   esters, 117–119, 148–149
   in wines, 115
Gallotannins, 118–119
Genistein, 107
Geraniin, 118
Glutathione, 99, 143–144, 169–172, 182
   addition reactions of, 170
   peroxidase, 171
   reaction with hydrogen peroxide, 170, 172
   reaction with hydroperoxides, 172
   reaction with hydroxyl radical, 171
   reaction with singlet oxygen, 171
   reaction with superoxide, 171–172

2″-O-Glycosylvitexin, 108
Guanine and guanosine
    oxidation products of, 9, 11, 42
    reaction with hydroxyl radical, 142
Gum guaiac, 2

H

Harmaline, 153
Harmalol, 153
Harmane, 153
Harmine, 153
Harmol, 153
Hesperetin, 103
Humic materials, 29
Hydrated electron, 28
Hydrocarbons
    autooxidation of, 30, 70
    in lipid peroxidation, 19
    polycyclic aromatic, 39–40
    reactions with hydroxyl radical, 46
Hydrogen peroxide, 10, 89
    reactions of, 44–45, 170
    see also Fenton reaction
Hydroperoxides (ROOH), 10, 32, 67, 78
    decomposition of, 32–33, 42, 59–61, 75
    determination of, 16
    from singlet oxygen reactions, 39–41
Hydroperoxyl radical (HOO·), see Superoxide
Hydroquinones, 26, 34, 84, 87, 122–126, 170
    reactions with singlet oxygen, 94
    see also Arbutin, Ubiquinol
Hydroxyl radical (HO·), 4, 25, 33, 38
    in carbohydrate oxidation, 11
    in DNA oxidation, 9
    in lipid peroxidation, 7
    in protein oxidation, 8
    reactions of, 45–46, 87, 96, 100, 109, 142, 145, 155, 171
Hydroxytyrosol, 128

I

Inhibitor, 73
Iron, see Transition metal ions
Isoflavonoids, 106–108
    see also Daidzein, Genistein, 2″-O-Glycosylvitexin
Isorosmanol, 131
Isovitexin, 103

K

Kadsurin, 119–120
Kaempferol, 89, 103
Ketones, see Carbonyl compounds
Kinetic chain length, 72

L

Lignans, 119–121
    see also Kadsurin, Nordihydroguaiaretic acid, Schisanhenol
Lipids, oxidation of, 5–8, 19–20, 41, 60, 70, 72, 98, 105
    see also Carbonyl compounds, Cholesterol, Conjugated double bonds, Hydrocarbons, Triacylglycerols
Lipoic acid, 174
Lutein, 183
Luteolin, 103

M

Maillard reactions, 12, 162
Malondialdehyde, 17
Malvin, 105
Melanin, 29
Membranes, see Cell membranes
Myricetin, 102

N

Naringenin, 103
Nasunin, 105
Neopterin, 162
Nicotine, 146, 155
Nitric oxide (NO), reaction with superoxide, 44
Nitro compounds, reaction with alkyl radicals, 76
Nordihydroguiaretic acid, 2, 120

O

Ovothiol, 173
Oxazoles, 162
Oxyberberine, 150
Oxygen
    ground-state, diradical, 25, 38

reaction with carbon-centered radicals, 31, 46
singlet, see Singlet oxygen
uptake, 15
Ozone, 8, 142

P

Peroxide determination, 15–17
Peroxides (ROOR), see Diperoxides
Peroxyl radicals (ROO·), 32–33, 46, 67
  in DNA oxidation, 11
  in lipid peroxidation, 7
  in protein and amino acid oxidation, 8
  rate constants for reactions, 35, 69–71
  reactions of, 33, 35–37, 70–71, 73, 76, 78, 85, 90–94, 96, 109, 114, 124, 126, 142, 145, 155, 181–182, 187
Petroleum, 4
9,10-Phenanthrenedione, 125
Phenol coupling, see Phenoxyl radicals
Phenolic acids, 110–115
  see also Caffeic acid, Ferulic acid, Gallic acid, Salicylic acid, Sinapic acid
Phenols
  acidity, 84
  reactions with hydroperoxides, 61
  reactions with hydroxyl radical, 46, 87
  reactions with peroxyl radicals, 70–71, 78, 83, 85, 90–92, 126
  reactions with singlet oxygen, 40
  reactions with superoxide, 43–44, 87–90
  redox potentials, 86–87
    see also Carvacrol, Catechins, Catechols, Flavonoids, Hydroquinones, Isoflavonoids, Purpurogallin, Resorcinols, Thymol, Vitamin E
Phenoxyl radicals, 83, 96
  dimerization (phenol coupling), 85
Pheophytins, 157–161
Phlogiston theory, 1
Phloroglucinol, 16
Photoionization, 27
Photosensitized oxidation, see Singlet oxygen
Phytochelatins, 57
Polyamines, 146
Polyethylene, 3
Polystyrene, 3
Polysulfides, 175
Preventive antioxidation, 51–52
Propolis, 117

Proteins, oxidation of, 8–9, 60, 97, 123
  see also Amino acids
Prunusols, 98
Purpurogallin, 127
Putrescine, 146

Q

Quercetin, 86, 93, 102, 105, 109, 114, 171
Quinones, 26, 34, 84, 90, 97, 105, 122–126
  reactions with alkyl radicals, 76–77, 125
  reactions with peroxyl radicals, 124
  reactions with phenoxyl radicals, 83
  reactions with superoxide, 99, 124

R

Radicals, see Free radicals
Redox potentials, 52–56
  of phenols, 86–87
Resorcinols, 129
Resorstatin, 129
Retarder, 73
Retinal, 186
Retinoic acid, 186
Retinol, see Vitamin A
Riboflavin, 89–90
Robinetin, 102
Rosmanol, 130
Rosmaric acid, 116
Rosmariquinone, 125
Rubber, 3
Russell tetroxide, 36
Rutin, 87, 94

S

Salicylic acid, 115
Schisanhenol, 119
Semiquinones, 26, 34, 84, 122, 124
Sesamol, 128
Sesamolin, 128
Sinapic acid, 114
Singlet oxygen ($^1O_2$), 38–42, 123
  in DNA oxidation, 11
  in lipid peroxidation, 7
  in photosensitized oxidation, 38–39
  in phototoxicity, 41
  physical quenching of, 58–59

reactions, 39–42, 92–94, 100, 109, 142, 171, 180, 183
Soot, 29
Spermidine, 146
Spermine, 146
Stobadine, 154–155
Strychnine, 146
Sugars, *see* Carbohydrates
Superoxide ($\cdot O_2^-$) and hydroperoxyl radical (HOO·), 12, 25, 28
    dismutation, 44–45
    reactions of, 42–44, 86–90, 99, 105, 124, 144, 150, 155, 171–172, 182
Synergism, in antioxidation, 61–62, 171, 187

T

Tannic acid, 118
Tannins, *see* Gallotannins
Tanshinone, 125
TBA (thiobarbituric acid)
    -reactive substances, 17
    test, 17–18
Tetrandrine, 149
Tetrapyrroles, 155–162
    see also Bilirubin, Chlorophylls, Pheophytins
Theaflavin, 109–110
Thiazoles, 162
Thiobarbituric acid, *see* TBA
Thiols
    reaction products of, with oxidants, 61, 75
    reactions with superoxide, 43, 61
    see also Glutathione
Thiyl radicals, 170, 182
Thymine, oxidation products of, 9
Thymol, 126
Tocopherols, *see* Vitamin E
Transferrin, 57
Transition metal ions
    as singlet oxygen quenchers, 59
    complexation of, 53, 55–58, 73–74, 102, 112, 121, 143, 170–172
    in ascorbate oxidation, 118–119
    in carbohydrate oxidation, 11, 13
    in free radical formation, 27, 30, 38, 45, 100
    in hydroperoxide decomposition, 33, 35, 59
    in lipid peroxidation, 7, 119
    in protein oxidation, 8
    in superoxide formation, 43, 77
Triacylglycerols ("triglycerides"), 6
Trolox, 109, 114, 127, 142

U

Ubiquinols, 97, 122
Ubiquinone, 124
Uracil, oxidation products of, 9
Uric acid, 141
    complexes with metal ions, 143
    oxidation products of, 142
    reaction with hydroxyl radical, 142
    reactions with peroxyl radicals, 142
    reaction with singlet oxygen, 142

V

Vinyl polymers, 29–30, 122
Violaxanthin, 184
Vitamin A, 186–188
    reactions with peroxyl radicals, 187
    synergism with Vitamin E, 187
Vitamin C, 13, 58, 98–100
    reaction with hydroxyl radical, 100
    reaction with peroxyl radicals, 71, 100
    reaction with phenoxyl radicals, 83
    reaction with singlet oxygen, 100
    reaction with superoxide, 43, 99
    synergism with Vitamin E, 62, 79–80, 97
Vitamin D, 14
Vitamin E, 13, 94–98, 122, 146, 185
    as singlet oxygen quencher, 59, 96
    pro-oxidant effects, 97
    reaction with hydroxyl radical, 96
    reaction with peroxyl radicals, 71, 79–80, 90, 96–97
    reaction with singlet oxygen, 92–93, 171
    reaction with superoxide, 88
    redox potentials, derivatives, 87
    synergism with glutathione, 171
    synergism with Vitamin A, 187
    synergism with Vitamin C, 62

X

Xanthophylls, *see* Carotenoids

Z

Zeaxanthin, 183–184